Lecture Notes in Morphogenesis

Series editor

Alessandro Sarti, CAMS Center for Mathematics, CNRS-EHESS, Paris, France
e-mail: alessandro.sarti@ehess.fr

More information about this series at http://www.springer.com/series/11247

Francesco La Mantia · Ignazio Licata
Pietro Perconti
Editors

Language in Complexity

The Emerging Meaning

Springer

Editors
Francesco La Mantia
Dipartimento di Scienze Umanistiche
Università di Palermo
Palermo
Italy

Pietro Perconti
Dipartimento di Scienze Cognitive
Università di Messina
Messina
Italy

Ignazio Licata
Institute for Scientific Methodology ISEM
c/o CNR
Palermo
Italy

ISSN 2195-1934 ISSN 2195-1942 (electronic)
Lecture Notes in Morphogenesis
ISBN 978-3-319-29481-0 ISBN 978-3-319-29483-4 (eBook)
DOI 10.1007/978-3-319-29483-4

Library of Congress Control Number: 2016940898

Printed on acid-free paper

This Springer imprint is published by Springer Nature
The registered company is Springer International Publishing AG Switzerland

To Franco Lo Piparo and Bernard Victorri

Contents

Introduction

Preliminary Questions

The book deals with devoted to three key questions concerning the relationship between complexity and natural language. Briefly, such questions are:

(a) What kind of complexity for natural language?
(b) Which theory of language in the perspective of complexity?
(c) What sorts of methods and models in the analysis of the observed phenomena?

All the essays in this volume show the reference to complexity as a constant element. However, the use of the singular may not be entirely appropriate. Let us see the reason why.

What Does "Complexity" Mean?

In the period between the end of the 1960s and the beginning of the new century, the word "complexity" was used to mean a variety of research programs aimed at different purposes, domains of investigation, and explanatory strategies; such as qualitative theory of dynamical systems, fractal geometry, physics of emergent processes, and cellular automata, to name a few. A slightly deeper analysis on this fleeting sketch would also show that, in each of the above-mentioned fields, the internal distinctions operating within the key-concepts are responsable for such big and significant differences. The concept of emergence offers, in this sense, an exemplary case: it was introduced to manage *novelty* and *unpredictability*, which are peculiar of the so-called «systemic properties», but it was later articulated in several other directions. In particular, we have to mention *computational* emergence to indicate those types of unpredictability produced by an imperfect knowledge of the initial conditions of a given system—but precisely for this reason «tame-able», i.e. compatible with the «possibility to compute step by step»—by a given model—the dynamics of the system. Instead, there has been talk of *observational* (or *intrinsic*) emergence to indicate those types of unpredictability that—far from being «tameable» in the specified sense—involve a «deviation» between the actual behavior of the system and the one described by the adopted model. Not only that: the two forms of emergences (for which, see Baas and Emmeche 1997; Licata 2005, 2008) admit a large variety of intermediate cases. Among other things, a certain ambiguity in the terminology has increased the misunderstandings that already proliferated around the concept. Thus some authors, such as Lichtenstein (2014), have used «computational emergence» to denote the patterns of interaction that are examined in the context of Artificial Life, while others, as Cariani (1992), used the expression «emergence-*relative-to-a-model*», instead of «observational emergence». In other words, this particular case (and many others, which we do not discuss here) clearly shows that "complexity" refers to a network of plurivocal and stratified concepts (such as the one of emergence). That's why some can think that the use of the singular "complexity" is not only inappropriate, but also incautious in epistemological terms.

How to justify our choice to put the singular name in the title? We believe that it is possible to respond by advancing two reasons: the former belongs to the semisecular debate around the repertoire of concepts that has been made available by the above-mentioned disciplines and the latter, on contrary, belongs to the debate that this book would like, to a certain extent, to inaugurate. Let us start from the first one.

Varieties of Complexity

According to the scholars in dynamical systems, fractals, cellular automata, etc., the word "complexity" represents a convenient umbrella term. The choice of the singular is firstly justified primarily by a practical reason: the ability to evoke synthetically a plurality of different research programs and well individuated on the theoretical level. But there is more: this choice not only is the outcome of a conventional agreement among the members of one or more scientific communities, but also the result of a careful reflection on the network of deep (and often unexpected) affinities that connect the disciplines and the involved concepts.

Just consider some few cases, it is worth remembering that a large portion of fractal geometry is an integral part of the qualitative theory of dynamical systems: the states of stability of certain classes of systems are instantiated from regions of the space of the phases that have fractal dimension—and which are called «strange attractors» (or «fractal attractors» *tout-court*). In addition, significant parts of the qualitative theory of dynamic systems occupy in their turn a leading role in the formal apparatus of the physics of the emergent processes: some peculiar properties of the attractors—of all the attractors, either strange or classical—are fundamental to describe/model emergent phenomena. Among these, the property of structural stability (Thom 1972; Stewart 1993; Devaney 1996) that captures, on the mathematical side, the «robustness» of these phenomena, i.e., their resistance (or «relative insensitivity») to local factors of perturbation (Emmeche et al. 2000). In the end, the theory of the cellular automata crosses, on different planes, each of these disciplines: it is a branch of the dynamic systems theory (since the cellular automata are special cases of *discrete dynamical systems*); a recent chapter of fractal geometry (since the patterns of interaction generated by cellular automata have fractal properties, as self-similarity); and an essential tool to simulate emergent collective behaviors (since the cellular automata are composed of *myopic* interacting algorithms). On the basis of this rapid reconnaissance, which, of course, has no claim to completeness, you can then better appreciate the value of the initial proposal: we use "complexity" (singular) not only (and not so much) to adhere to terminological choices of a certain scientific community, but also (and especially) to highlight the presence of a dense network of links between some relevant portions of the mathematics of the twentieth and the twenty-first centuries that have never ceased to interact and interface. Based on this observation, the step that leads us to expose the second reason is very short.

Complexity in Linguistics

One of the main purposes of this work is to examine the impact of some types of mathematics on disciplines related to the sciences of natural language: in the first place, theoretical linguistics, but also semiotics, the discourse analysis and the vast spectrum of works, located halfway between logic and philosophy of language, which have as their object theoretical problems of capital importance, such as the statute of the compositionality and lexical polysemy. In these theoretical contexts, the reference to "complexity" is equivalent to defining, with a short margin of uncertainty, the identity of these particular types and their role in the areas mentioned above. The use of the singular number has a twofold strategic value for us: not only is it in continuity with the idea, strongly shared by who here write that the mathematics, we deal with, makes up an *archipelago* of formal interconnected entities. But above all, it is consistent with the purpose of identifying a mathematical universe that, even in the variety of its aspects, has peculiar and rather homogeneous features, and whose repercussions in the analysis of natural language constitute a unique event in terms of epistemology of models and, more particularly, of the *naturalization* of meaning (see at least, Petitot et al. 1999). To justify this uniqueness, we have, to recall, through a brief overview, some crucial moments in the history of ideas and linguistic theories, with particular attention to the relationship established in the last 60 years between theoretical linguistics and mathematics.

Background: Computational Grammars and Generativity

The Chomskyan project of the foundation of a theory of computational grammars (see at least, Chomsky 1957; Greibach 1978) is one of the main epicenters. Born as an offshoot of Harrisian mathematical linguistics (see at least, Harris 1951), from which however it soon significantly differentiated (see Tomalin 2006, 110–111), it imported large slices of discrete mathematics in some privileged domains (theory of computability, graph theory, portions of the theory of algebraic structures, etc.) into linguistic analysis; essentially: morphology and syntax. The most profound results are at least three:

- The institution of a *hierarchy* of abstract computing devices (or «Chomsky's Hierarchy»), designed to specify, namely «generate», certain classes of formal languages sorted by the growing (and/or descending) generality of their expressive power.
- The use of such languages in describing/modeling of peculiar aspects of the natural language and, in particular, of some languages. For instance, the use of «regular languages» in the description of the morphology of the English language.
- The identification of the peculiar generativity of the historical-natural languages with the generativity of formal languages and (in a broad sense) with the computational automatisms of abstract devices designed to specify («computational grammars») or «recognize» («automata») these languages.

Especially this last point has caused a computational interpretation of the concept of grammatical rule (i.e., the grammar rules as «rewriting rules»)[1] and a recursive interpretation of syntax (i.e., syntax as a system that admits the possibility of «unlimited *embeddings*»). Therefore, the program of naturalization of the meaning, arisen at the intersection between the qualitative theory of dynamical systems and physics of emergent processes—just to mention the main *nodes* of the aforesaid affinity network—has among its initial objectives the setting of a radical critique of the idea of language subtended by computational grammars. Known under the more disparate labels («Semiophysics», in Thom's lexicon 1988 and Thom 1990; «Morphogenesis of meaning» or «Physics of meaning», in the respective lexicons of Petitot 1985 and Petitot 1992), it has individuated in the Chomskyan notion of generativity, one of its principal polemical targets. The synthetic expositions of the main stages of this attack allow you to trace, in our opinion, certain features of the scientific undertaking that has inspired (even in disagreement) the general trends of some of the most representative works of this program.

[1]For a critical reconstruction of this crucial phase, see among others, Auroux (2009, pp. 46–47).

Critique of the Generativist Reason: Thom and Petitot's Semiophysics

Thom and Petitot's *Semiophysics* (or *Physics of meaning*) promotes a radical critique of the idea of language subtended by the Chomskyan computational grammars through the revision of some fundamental theoretical instances. The second one involves at least three stages:

(a) Focusing on the irreducibly peculiar character of the generativity of historical-natural languages (Thom 1979; Petitot 1982, 1985; Bathelemy et al. 1996);
(b) Identifying a «limited recursion» operating in the syntax of historical-natural languages (Petitot 1982, 1985; Barthelemy et al. 1996).
(c) Individuating the empirical characteristics of the «linguistic datum»—or its «phenomenology» (Petitot 1982, 39) and expelling the recursion from the repertoire of formal invariants attributed to these syntaxes (Petitot 1982; Barthelemy et al. 1996).

Limited Recursion and Naturalization of the Meaning

Briefly, these are the starting points of the main program. It is difficult to establish its theoretical value. In fact, these are critical observation which concern a precise historical moment of the Chomskyanism—i.e., the twenty-year period between the end of the 1950s and the end of the 1970s. Located in a long period of time—that includes subsequent revisions of the *Chomskyan generativism* (then *minimalism*) and the objections to this from the outside (i.e., from paradigms as the *connectionism*[2] and *neo-connectionism*[3]) and from the inside (i.e., from heterodoxies more or less marked as the *biolinguistics based on epigenesis*[4]); they appear rather dated or generic. In addition, if one only considers the third point, these observations seem to be definitely controversial: the most recent and hot discussions about the status of universal linguistics (Christiansen et al. 2010) include the recursion between the species-specific formal properties of the human language faculty, while the discovery of historical-natural languages different from recursive constraint (Everett 2005; Everett 2009) has been greatly reduced in the context of subsequent works (Bickerton 2009 and, in a different perspective, Zwart 2011). Indeed, there are reasons to believe that the program of naturalization of meaning started by Thom and Petitot has rested since its very beginning on very fragile foundations.

[2]Elman et al. (1996).

[3]Christiansen and McDonald (2009).

[4]Longa and Lorenzo (2003), Longa (2008), Boeckx and Longa (2011).

However, things are not precisely in this way. Indeed, in the context of historical reconstruction of the problem, we can say that they are exactly in the opposite way. For instance, if in linguistics the reference to a limited recursion may seem almost obviously granted (Christiansen 1992; Christiansen and Chater 1999; Folia et al. 2011), it should be not forgotten, however, that it is the result of a slow and complicated process of conceptual transformation. In the domain of computational grammars—and, more generally, in the whole field of action of Chomsky's reflection—the possibility of a limited recursion is accepted only as a by-product of neurophysiological constraints (such as the «finite amount of memory») that operate on the mechanisms of production/understanding of the speaker. The semiophysics aims directly at the heart of this theory in order to reverse, so to speak, the epistemological perspective from which it had originated. The reversal consists in a substantial requalification of the observed empirical limitations: the neurophysiological constraints (or *execution* bonds) assume the status of a «critical regularity», i.e., of linguistic experiences that deny systematically the hypothesis of an ideal syntax (or *competence*) compatible with the possibility of unlimited *embeddings*—which nobody has ever observed! (See Victorri 1997). In the context of this requalification, limited recursion is the peculiar manifestation form of such regularity. It acquires the role of «observable», i.e., a linguistic phenomenon that, far from being indexed under the heading of negligible differences—requires an explanation as well as a formal and appropriate analysis:

C'est un piètre subterfuge que de rejeter—comme le fait Chomsky—la responsabilité des inadéquations du modèle formel sur l'incapacité d'une «performance» à assumer les capacités génératives d'une infatigable «competence». [...] *C'est l'autolimitation des capacités génératives de la syntaxe qui demande explication.*[5]

The point (a), from which the main consequences outlined in (b) and (c) derive, offers, if examined in detail, a good ground for the required explanations.

Rethinking the Generativity of Languages from the Use

The generativity of natural languages is dissimilar from generativity of formal languages in the sense that the assembly of elementary units in more extended composite units (peculiar operation of any syntactical combination) exhibits, in the two cases, deep differences. The most strident difference, according to Thom and Petitot, affects the sphere of the meaning—or, with terminology oriented to the pragmatic, «the sphere of the use». The combinatorial syntactic structure of natural languages provides composition rules which are "sensitive" to the actual uses of the produced expressions. The iteration of these rules is thus bound to the empirical limitations of speakers and (in a complementary way) to the conditions of

[5]Thom (1980, p. 164). My italics.

intelligibility (or acceptability) determined by discourse situations. On the contrary, the combinatorial syntactic structure of formal languages are based on composition rules (concatenation and derivation) that operate in abstract mode, i.e., «independently from» the empirical limitations of speakers. Therefore, the latter model of generativity differs radically from the former. Its importation in the linguistic field is not only inappropriate—because it cancels substantial and irrepressible diversities—but especially unrealistic because it determines a detachment between the plane of the syntax and the plane of the semantics (or pragmatics) which finds no empirical correspondence. In a natural language that involves as combinatorial syntactic structure, a set of iterable composition rules independently from execution constraints produces in fact complex expressions of infinite length, i.e., phrases with an unlimited number of embedded subordinates—and therefore unusable and/or incomprehensible! So:

[…] les linguistes "formalistes" se sont efforcés d'axiomatiser la grammaire et la syntaxe. Là, ils ont dégagé un certain nombre de processus formels, les grammaires génératives et transformationnelles, dont la validité, sur le plan d'une description formelle des phrases effectivement présentes dans le corpus—ne peut être niée. *Mais, si l'on systématise ces processus formels en une axiomatique dont on poursuit aveuglément la complétion formelle, on ne tarde pas à construire des phrases d'une telle longueur et d'une telle complexité qu'elles en perdent toute intelligibilité.*[6]

More precisely:

L'école des grammariens générativistes a cru pouvoir douer la syntaxe de propriétés génératives indéfinies, c'est à dire, permettre à la grammaire d'engendrer des phrases comportant une infinité de subordonnées emboîtées. *C'est là, inutile de le dire, une pure fiction, la très concrete performance vient freiner l'élan génératif de la compétence, en sorte que ce procédés itératifs s'arrêtent eux-mêmes au bout de quelques opérations (quatre ou cinqu, au plus).*[7]

On the basis of these insights, «the auto-limitation of generative capacity of the syntax» is explained in reference to the conditions of actual intelligibility (i.e., use and understanding) that any produced form must meet in order to circulate in discourse situations: «à vouloir maintenir dans la langue une générativité sans frein, l'expression formelle perd tout contact avec son signifié, en fait, perd tout sens».[8] The reference to a limited recursion (b) and the elimination of the recursion (in the sense of an *infinite* recursion[9]) from the repertory of the formal invariants attributed to natural languages (c) are both strategic moves that belong to the order of ideas just outlined.

[6] Thom (1970, p. 13). My italics.

[7] Thom (1980, p. 294). My italics.

[8] Ibidem p. 292.

[9] Cf. Petitot (1979).

Oscillations and Gaps

These are the results that Thom and Petitot achieved. The recent history of reflections elaborated within the semiophysics (or the works it inspired) confirms the Chomskyan generativity model in its role of privileged polemical target. The concept of recursion has hence remained at the center of successive controversies having all the same problematic point: the sketch of a realistic critique of the idea of language suggested by computational grammars and, in a broad sense, by the Chomskyan approach (see at least, Wildgen 2009). However, there have been dissonant voices in the «chorus», as Victorri (2005, 2007), which have contributed to a partial rehabilitation if not of the Chomskyan undertaking—respect to which a very strong general dissent still remains—at least of the recursion. The partial recovery of the concept began from syntactic phenomena of syntactic subordination. In this context, the recursion was presented as a computational device of the faculty of language that ensures a nontrivial property for planning the grammar of the discourse, namely the embedding of sentences:

[...] [...] la récursivité joue un rôle fondamental dans la complexité des énoncés. C'est la récursivité qui est à l'oeuvre quand on enchâsse une proposition dans une autre, qui, elle-même peut déjà être enchâssée, et ainsi de suite, sans qu'il y ait de limite au nombre des fois que l'on peut réitérer cette opération.[10]

However, elsewhere, the attenuation of pioneering criticism of Thom and Petitot seems to coexist with a cautious skepticism that strongly repeats the idea of a limited recursion. Paradoxically, the phenomenon of syntactic subordination appears in the discussed studies to justify the rejection of an unlimited recursion:

Mais ce n'est pas pour autant forcément le mécanisme de récursivité infini, à l'oeuvre à tous les niveaux de l'organisation syntaxique, postulé par la grammaire générative chomskyenne. En fait, on peut se passer de la récursivité chomskyenne pour le traitement des syntagmes nominaux et prépositionnels, pour l'organisation actancielle, etc., et, de plus, là où ce mécanisme est vraiment nécessaire, c'est-à-dire pour l'enchâssement d'une proposition à l'intérieur d'une autre, le nombre de niveaux hiérarchiques à traiter de manière récursive est toujours très faible.[11]

The justification for this rejection is an important evidence that confirms the general trends of a whole research program. Therefore, despite some signs of partial rapprochement, the distance between the two programs, though in the wide variety (sometimes indefinite) of positions taken by individual authors, remains deep.

[10]Victorri (2005, p. 12).

[11]Victorri (2009, p. 55).

Syntax in Semiophysics

Behind this gap, there is, in our opinion, a profound reason that deals with the status of the syntax—and, more generally, with the relationship between syntax, semantics, and pragmatics. According to the authors inspired by the program of naturalization of meaning started by *Semiophysics*, there is no solution of continuity between these three key aspects of the natural language: syntax, semantics, and pragmatics are inseparable levels of organization. The reference to limited recursion (and its rehabilitation in terms of peculiar regularity of the linguistic experience) clarifies the shape of this bond: the number of embeddings found in complex expressions of a given language is limited—and, indeed, reducible to a few levels of «depth»—since the combinatorial syntax of natural languages are constrained by the actual conditions of intelligibility (i.e., the use and understanding) that every expression must meet in order to circulate within real or imaginary discourse situations. The recursion of natural languages is therefore limited because it is a property of syntax «usage based»—or, as Petitot suggests (1985: 131): «une récursivité à courte portée rends la sémantique *inseparable* de la syntaxe».[12] Starting from this observation, which has become in recent years the theoretical cornerstone of innovative approaches fundamental to the study of the linguistic recursion, including the so-called usage-based approaches (see at least, Christiansen and McDonald 2009), the studies inspired by semiophysics have developed a vision of the relationship between syntax and semantics that presents two complementary elements of radical novelty:

1. The semantics as «organizing centre» of the syntax, i.e., as its «principle of implicit production»—and not as its «interpretation»;
2. The syntax as «result» of the semantics, i.e., as a «by-product of the construction of meaning»[13] (i.e., a semantics that is «inseparable» from pragmatic factors)—and not as a «guide» /«starting point» of the interpretation (or «attribution of content»).

The two hypotheses have considerably reduced the Chomskyan principle of autonomy of the syntax—and its more sophisticated versions that provide the possibility of «levels of interface» between syntax and semantics (see at least, Partee 2014). In fact, the matter is not to «map» a component in the other (see., among others, Legendre et al. 2001), i.e., to establish a collection of oriented assignments from syntax to semantics (i.e., the semantics as «interpretation» of the syntax) or from semantics to syntax (i.e., the syntax as a «guide» of the semantical interpretation), but to interpret the configurations of forms exchanged between the interlocutors as structures governed by «constraints» that emerge from global interactions (or dynamics) finely distributed between the semantic endowments of each local form and that instantiate, at certain levels of description, the topologies of

[12] Italics in the text.

[13] Victorri (2000, p. 17).

complex attractors which in turn can decompose (or «parcel themselves out») in many other «sub-attractors»—one for each form or constituent. As Petitot stated (1985, 148): «[...] on considérera qu'une phrase est un signifié *global* décrit [...] par la topologie d'un attracteur complexe d'une dynamique [...] supposée à titre purement implicite. [...] La description structurale de la phrase [...] exprime le processus de production à travers lequel cet attracteur *ce morcelle* [...] en sous-attracteurs par une attribution lexicale les constituants immédiats».[14]

Significant Intersections: Grammars and Cognitive Semantics

However, the decrease of the autonomy of the syntax, a common trait generativism but also to a good part of the classical cognitive science, is not the exclusive prerogative of Semiophysics. In the first half of the 1990s, numerous research programs, as cognitive grammar (see at least, Langacker 1994) and cognitive semantics (see at least, Talmy 2000), have followed similar paths. Even for these approaches the autonomy of syntax has represented a «myth» to demolish. This caused the proliferation of work oriented, on one hand, towards a perceptual (or experiential) *grounding* of the syntax; on the other hand, in the direction of a primacy of the semantics—by which, of course, the inseparability of «grammar and meaning» is justified.

Furthermore, one should not overlook the important exceptions within the minimalist galaxies of the last generation. For instance, the «Simpler Syntax» (see Culicover and Jackendoff 2005) has proposed a model of levels of interface between syntax and semantics based on the federation principles of cognitive grammar promoting the idea of a constant interaction (or *interplay*) «between syntactic phenomena and details of the lexical meaning».[15]

Toward a «complex» Science of the Natural Language: the Semiophysics' Inheritance

Nevertheless, beyond the similarities and the single exception which attest the presence of common objectives and problems, the two principal assumptions of semiophysics maintain, in our opinion, a significant part of originality. Thanks to a useful comparison with the linguistics of enunciation—and, in particular, with Antoine Culioli's Theory of Predicative and Enunciative Operations (henceforth, TOPE)—semiophysics has managed to draw a set of fundamental theoretical

[14]Petitot 1979. For more information, see at least Petitot (1995, 2014).

[15]Culicover and Jackendoff (2005, p. 534).

consequences for subsequent developments of new research programs that have influenced the international semiotic language debate in the last decade. We are referring to the contributions of the team of Victorri and Fuchs (1996), Victorri (2004), Visetti and Cadiot's Theory of Semantic Forms, (2001, 2006) or certain morpho-dynamical and phenomenological readings of linguistic structuralism proposed by Piotrowski (1997, 2009). Different for purposes and objectives, these works are currently one of the major epicenters of the studies on the natural language inspired by the mathematics of complexity. In particular, the researches carried out within the framework of these programs are centered on some key issues of semiotic reflection. Let us look at these briefly.

Problems Inspired by a Complex Science of the Natural Language

The issues at stake are essentially three as follows:

(a_1) Status of the compositionality with particular reference to the phenomena of lexical polysemy (Victorri and Fuchs 1996; Visetti 2004a);

(b_1) Status of the lexical units with particular reference to the enunciative construction of meaning (Victorri and Fuchs 1996)—and, in a broader sense, to the dialogical conditions of emergence of the relationship between the expression and the content plan (or *semio-genesis*, Piotrowski and Visetti 2014);

(c_1) Status of mathematics with particular reference to the description of semio-linguistic phenomena (Victorri and Fuchs 1996; Visetti 2004b; Piotrowski and Visetti 2014).

Gestalt Compositionality

With reference to the point (a_1): the primary intuition of the compositionality is adapted to the co-textual (and contextual) changes of lexical meaning. In other words, the link between the meaning of an utterance and the meaning of the lexical constituents is compositional, but, in the context of the proposed adaptation, is integrated with a link that certain authors, as Rastier (2001), called «decompositional» and that reflects, in fact, the Fregean Principle of the Context. Based on this integration, the enunciative meaning is determined by lexical meanings in turn determined by the enunciative meaning. The circularity of the generative mechanism is presented as the only plausible form of compositionality since it implements a system of reciprocal determinations (i.e., oriented from the parts to the whole and from the whole to the parts) that keeps together two physiological complementary aspects of the language activity: the meaning of an entire utterance rests on the meaning of its parts and the meanings of the parts is context-sensitive. Hence the idea of a «Gestalt compositionality»: «<It is necessary> to consider the utterance as a «Gestalt» where the relationship between the whole and the parts are absolutely bidirectional».[16]

Lexical Polysemy and Semiogenesis

The point (b_1) examines deeply some of the consequences arising from the adoption of a Gestaltist point of view on the compositionality. The contribution offered by the lexical units in the building of enunciative meaning (or the semiogenesis) is an «undifferentiated amount of signification» subject to Gestalt compositionality. Less metaphorically, the contribution of the sub-enunciative constituents (and in particular of the lexical units) may not be fixed before the interactive constraints of the enunciation come into play. In addition, since the bonds are in the form of reciprocal determinations «parts-whole/whole-parts», the contribution must be sufficiently indeterminate or «flexible» to ensure the *local* emergence of the lexical content consistent with the provisional «meaning unities» which are produced. During the last 20 years, the authors engaged in this kind of analysis have proposed to designate the genetic matrix of such a content with various and several labels. Some of them spoke of «semantic potential» (Victorri 1992; Victorri and Fuchs 1996)—using a phrase introduced in the literature by Halliday (1978), but already present in the lexicon of the second French phenomenology (Ricoeur 1975) and then subsequently in some recent works of cognitive semantics (Allwood 2003; Evans 2009). And others spoke of «schematic forms» (see Col et al. 2010)—inspired by the metalinguistic lexicon of TOPE (Culioli: 1986); or even «motif» (Cadiot and Visetti 2001)—orienting in the direction of phenomenological tradition (Husserl, Gurwirtch, Merleau-Ponty) or Gestalt (Koffka, Köhler, Wertheimer) and in the direction of the philosophy of symbolic forms of Cassirer. If you omit the variety of positions which may be found behind such a richness of terminological references, moreover at source of very productive «perspectival conflicts», you can find a common general orientation intent on identifying a particular aspect of the lexical units.

Potential semantics, *schematic forms*, or *motifs* are all theoretical constructs made to capture the availability (or capacity) of lexical forms to interact with other forms and to produce (by virtue of these interactions) the «meaning-unities» (utterances, discursive fragments, etc.) mediated by semiogenesis. Hence a «praxeological» conception of the lexical units oriented to the individuation in each word not so much (and not only) of the carrier of certain semantic endowments but of the activator of the semiotic processes (or «sémiogenies») peculiar to language activity, including the union of local forms in «meaning unities» that emerge in discourse situation.

[16]Victorri (1992, p. 242).

Local Models and Qualitative Mathematics of the Continuous

The point (c_1) is inscribed in this wide set of problems. For the research programs, which have arisen between semiophysics and TOPE, the use of mathematical tools is justified by a precise overriding objective—and that is the elaboration of metalinguistic representations adhering to the examined language phenomena. Therefore the analysis conducted in this domain of investigation appears in close continuity with the analysis of the generativist school. In particular, they converge, in our opinion, at least on one point: the *mathematization* of the semio-linguistic structures. For both scholars of Chomskyan orientation and for scholars of semiophysics and TOPE the application of mathematical concepts to the study of linguistic phenomena is the condicio sine qua non of the naturalization of the meaning, i.e., of the integration of metalinguistic representations in the vocabulary of the physical-natural sciences. Compared to the path taken by Chomskyanism, the process of mathematization started by the epigones of Thom, Petitot, and Culioli diverges however, radically on two fundamental aspects:

(1) the diversity of natural languages;
(2) types of mathematics on which to base the elaboration of metalinguistic representations.

Concerning the point (1), the detachment with Chomskyan formulation is clear: the formal apparatus of natural languages—far from being reduced to epiphenomenal variations of internal states of the language faculty (see at least Chomsky 2000)—are considered as bearers of peculiar differences that require the adoption of appropriate metalinguistic tools, i.e., «sensitive» to the identified differences. Therefore, the matter is not to encode the diversity of languages in terms such as «parameters» (or «points of choice») provided by Universal Grammar (see Rizzi 1991), but to fabricate «local models», i.e., computing devices capable of detecting on a metalinguistic plan rules for specific languages—and not principles genetically induced by the faculty of language (Culioli and Desclés 1981; Culioli 2002). The difference is substantial and has important implications both on the choice of the objects of inquiry and on the methods adopted to define them. For instance, it involves an epistemological rehabilitation of the concept of *grammar rule*. According to the supporters of the minimalist shift, a grammar rule is only a «taxonomic expedient» (Chomsky 2000), which is a useful fiction within the limits of informal descriptions of linguistic phenomena, but entirely without theoretical value for the purposes of rigorous formal descriptions. On the contrary, for the authors mentioned above, grammar rules have theoretical dignity for the fact that they are the main access to configurations of forms (or *markers*, in the lexicon of TOPE) which are trace of underlying language operations (see Culioli 2000; Desclés 1990). The task of simulating on metalinguistic plan the «form» of these operations and of testing the intertranslability, i.e., the ability to be «transferred» (out of possible variations) from a model to another belongs to the local models. This implies an ideal of mathematization of the semio-linguistic structures that represents, in relation to the Chomskyan approach, a first novelty on the status of the inter-linguistic invariance: the language invariants are neither abstract characteristics of the initial state of the language faculty (or *formal universals*) nor «fixed elements that enter in special grammars [languages] (or *substantive universal*)», but sets of operations extracted from the formal structure of natural languages and examined through the metalinguistic «filter» of local models that attempt to verify the intertranslability. But this is not all. There is at least another novelty that marks a fundamental «gap» between the two programs and affects the types of mathematics used in the construction of metalinguistic representations, which of course leads to the point (2).

If the project of founding a theory of computational grammars—and, more generally, the Chomskyan approach—is characterized by the use of large portions of the discrete mathematics (see Sect. 2.1), then the qualitative mathematics of the continuum (geometry and differential topology, and dynamical systems theory, etc.) is, instead, the main reference point for the most of the works inspired by Thom, Petitot, and Culioli. They are already present in the foundational texts of semiophysics and TOPE (albeit, in the second case, in smaller proportions) and they have been applied massively to the analysis of the lexical polysemy and semiogenesis. In particular, the main results of these applications develop some insights on modeling of the *context-sensitive* aspects of lexical meaning largely anticipated by Petitot (1985) and Culioli (1986).

The key concept, roughly, is: the meanings generated by a lexical unit in the enunciative plans correspond to the states of local equilibrium of a system changing in time. This system is given by the interaction patterns of the units, the co-textual indexes, and the situation of enunciation. Every meaning generated thus constitutes a stability state of these interactions, i.e., a state of local equilibrium on which the contextually relevant interpretations of the examined unit will stop. From here we have, the construction of metalinguistic representations based on the conceptual repertoire of qualitative theory of (continuum) dynamical systems. In the context of these representations, each meaning generated by the lexical unit is described as the «attractor» of interpretations contextually relevant. The analogy is powerful and hides the idea that every lexical meaning behaves like a powerful magnet that «attracts» interpretations of the units that are compatible with the enunciative plans in which it occurs. More precisely—but eliminating the technical terms (see Victorri and Fuchs 1996; Salanskis 2001): every interpretation is graphically represented by a path (or relief) on a continuous space S that instantiates the totality of the word's possible meanings. S is called «semantic space» and is divided into several subspaces or regions that define the neighborhoods (i.e., the basins of attraction) of one or more possible attractors. Within these regions, the graph of the paths (or dynamic) that describe the contextually relevant interpretations of the unit bends itself downward, i.e., has some points of «low altitude» which define the «attractors» of the semantic space. To summarize, if compared to the naturalization program of meaning started by Chomskyanism, the work inspired by the semiophysics and TOPE are different for at least two irreducibly peculiar aspects: the statute assigned to the diversity of languages and types of mathematics adopted in the construction of metalinguistic representations. The reference to complexity is stated so in a twofold sense: (a) «complexity» as «heterogeneity» of examined phenomena and «sensitivity» of the models adopted to describe them; (b) «complexity» as «co-presence» of complementary phenomena, i.e., stability of semantic attractors (or interpretations of contextually relevant) and continue variability of the meaning.

Conclusion

The analysis conducted on some of the key issues of the relationship between complexity and natural language has allowed us to focus on the main stages of the naturalization program of the meaning which is, for methods and purposes, alternative to the Chomskyan generativism (then minimalism). Here is a brief summary:

(a) Justification for the irreducibility of peculiar generativity of natural languages rather than the generativity of formal languages in the Hierarchy of Chomsky;
(b) Identification of a limited recursion operating in the syntax of the natural languages;
(c) A radical rethinking of the Chomskian principle of autonomy of the syntax;
(d) Gestalt compositionality;
(e) A praxeological conception of the lexical units;
(f) Theoretical rehabilitation of the notion of grammar rule;
(g) Introduction of local models sensitive to the diversity of languages;
(h) Construction of metalinguistic representations based on mathematics of the continuum.

These, briefly, are the crucial points of a theoretical work conducted over nearly half a century by teams of scholars mainly operating in France and in some important research centers in Europe—as the Danish school of Bungaard (2010), Brandt (2006) and Ostergaard (1997), or the German school of the already mentioned Wildgen (2002).

The essays here collected try to «sketch out», in a peculiar way, the issues we tried to expose in the course of *this* introduction. In particular, the first part of the book (Complexity and linguistic theory: epistemological questions) include four contributions devoted to the analysis of the epistemological statute of the mathematics of complexity (Piotrowski, Visetti), the role of mathematical continuum in modeling certain semio-linguistic phenomena (Poibeau, Col, De Angelis), the dynamics of attractors in the study of linguistic change and grammaticalization (Fagard, Nadal, Feltgen), and the quest for the condition of cognitive plausibility which computational psychology should satisfy in order to be compatible with the naturalistic turn in the cognitive science (Perconti). The second part (Complexity, semiotics and enunciation theory) is devoted to the examination of some particular study cases inspired by the mathematical complexity in the field of semiotic analysis and the theory of enunciation (Paolucci, Bondì, La Mantia). The third part of the volume (Linguistic Complexity: physics, computation and biology) investigates the opportunities of quantum physics in modeling linguistic operations (Arecchi), how to bring traditional computational and complexity approaches together (Zeppi), and how to develop neuro-computational models for language and brain complexity (Plebe and De La Cruz). All the following essays are dealing with some of the most intriguing issues in the current science of the mind, hoping that they will produce new reflections on this topic.

References

Auroux, S. (2009). «Mathématisation de la linguistique et nature du langage». *Histoire, Epistémologie Langage*, *31*(1)19–59.
Allwood, J. (2003). «Meaning potentials : some consequences for the anlysis of variation in meaning. In H. Cuyckens et al., *Cognitive Approach to Lexical Semantics* (85–125). Berlin-New-York: Mouton De Gruyter».
Baas, N., & Emmeche, C. (1997). «On emergence and explanation». *Intellectica*, *2*(25), 67–83.
Barthélemy, J. P., et al. (1996). «Logique et dynamique de la cognition». *Intellectica*, *2*(23), 219–301.
Bickerton, D. (2009). «Recursion: core of complexity or artifact of analysis ». In T. Givón & S. Masayoshi (Eds.), *Syntactic Complexity: Diachrony, acquisition, neuro-cognition, evolution,* (pp. 531–544).
Cadiot, P., & Visetti, Y. -M. (2001). *Pour une théorie des formes sémantiques*. Paris: P.U.F.
Culioli, A. (2000). *Pour une linguistique de l'énonciation*. Paris: 3 Tomes, Ophrys.
Culioli, A. (2002). *Variations sur la linguistique*. Paris: Klincsieck.
Culioli, A. (2014). *L'arco e la freccia*, a cura di Francesco La Mantia, Prefazione di Tullio De Mauro. Bologna: Il Mulino.
Culioli, A., & Desclés, J. P., *Systèmes de représentations métalinguistiques*. Paris: Ophrys.
Cariani, P. (1992). «Emergence and artificial life», In C.G. Langton, C. Taylor, J.D. Farmer & Rasmussen S. (Eds.), *Artificial life II* (pp. 775–798) Sante Fe Institute Studies in Sciences of Complexity, v. X. Addison-Wesley: Reading, MA.
Chomsky, N. (1957). *Syntactic Structures*. Cambridge, Massachussets: M.I.T Press.
Chomsky, N. (2000). *New horizons in the study of language and mind*. Cambridge (Mass.): MIT Press.
Christiansen, M. (1992). «The (non) necessity of recursion in natural language processing». In *14th Annual Cognitive Science Society Conference* (pp. 8–22). Bloomington: Indiana University
Christiansen, M. (1999). «Towards a connectionnist model of recursion in human languistic performance». *Cognitive Science*, *23*(2)157–205.
Christiansen, M., & Chater, N. (1992). «Finite models of infinite language : A connectionnist approach to recursion». In M.H. Christiansen & N. Chater (Eds.), *Connectionist psycholinguistics* (pp. 138–176). Westport.
Christiansen, M., & McDonald, M. C. (2009). «A Usage-based approach to Recursion in Sentence-Processing». *Language Learning*, 59(1)126–161.
Culicover, P., & Jackendoff, R. (2005). *Simpler Syntax*. New-York: Oxford University Press.
Desclés, J. P. (1990). *Langages applicatifs, langues naturelles, cognition*. Paris: Hermés.
Devaney, R. (1996). *Chaos, fractals, and Dynamics*. Milano: Addison-Wesley.
Ellman, J. et al. (1996). *Rethinking innatennes*. Cambridge: Cambridge University Press.

Emmeche, C. et al. (2000). «Levels, emergence and three versions of downward causation». In P. B. Andersen, C. Emmeche, N. O. Finnemann & P. Voetmann Christiansen (Eds.), (2000): *Downward Causation. Minds, Bodies, and Matter*. Århus: Aarhus University Press.
Evans, V. (2009). *How words mean*. New-York: Oxford University Press.
Everett, D. (2005). «Cultural Constraints on Grammar and Cognition in Pirahã». *Current Anthropology*, *46*(4) 621–646.
Everett, D. (2009). «Pirahã Culture and Grammar: A Response to Some Criticisms». *Language*, *85*(2) 405-442.
Folia, V., et al. (2011). «Implicit Artificial Syntax Processing : Genes, Preferences and Bounded Recursion». *Biolinguistics*, *5*(1–2)105–132.
Greibach, S. A. (1978). «Comments on universal and left universal grammars, context-sensitive languages, and context-free grammar forms». *Information and Control*, *39*135–142.
Halliday, M. (1978). *Language as Social Semiotic*. London: University Park Press.
Harris, Z. (1951). *Linguistics. Structural Methods in linguistics*. Chicago: University of Chicago Press.
Legendre, G. et al. (2001), *Optimality-Syntax theory*, MIT PRESS, Cambridge, Massachussets.
Licata, I. (2005). «Oracoli di Turing e sistemi logicamente aperti». *Intelligenza artificiale*, *62*41–43.
Licata, I. (2008). «Logical Openness in Cognitive Models», in *Epistemologia*, *31*(2)177–192.
Lichtenstein, B. (2014). *Generative Emergence : New Discipline for Organizational and social innovation*, Oxford University Press, 2014.
Langacker, R. (1994). *Foundations of Cognitive Grammar*, I-II, Stanford University Press.
Legendre, G. et al., *Optymality-theoretic Syntax*, Cambridge Massachussets: MIT Press.
Longa, V., & Lorenzo, G. (2003). «Minimizing the genes for grammar. The minimalist program as a biological framework for the study of language». *Lingua*, *113*643–657.
Longa, V., & Lorenzo, G. (2008). «What about a (really) minimalist theory of language acquisition?». *Linguistics*. 46(3)541–570.
Longa, V., & Lorenzo, G. (2009). «Beyond generative geneticism: Rethinking language acquisition from a developmentalist point of view». *Lingua*, *119*1300–1315.
Longa, V. (2008). «Una visión crítica sobre la noción de 'programa genético' desde la biología y la lingüística: consecuencias para la conceptualización de la ontogenia del lenguaje». *Verba*, *35*347–385.
Partee, B. (2011). «Teaching formal semantics». *Teaching Linguistics: Reflections on Practice*, ed. Koenraad Kuiper, pp. 40–50, London: Equinox Publishing.
Petitot, J. (1982). «Sur la signification linguistique de la théorie des catastrophes». *Mathématiques et Sciences Humaines*, *79*37–74.
Petitot, J. (1985). *Morfogenèse du sens*. Paris: PUF.
Petitot, J. (1992). *Physique du sens*. Paris: éditions du C.N.R.S.
Petitot, J. (1995). «Morphodynamics and attractor syntax». In T. van Gelder & R. Port (Eds.), *Mind as Motion : Exporations in the Dynamics of Cognition* (pp. 227–283). MIT Press
Petitot, J. (2014). «Aspects du structuralisme dynamique». In F. Lo Piparo, F. La Mantia, & C. Paolucci (a cura di) *Semiotica e matematiche* (pp. 11–61) Versus, Quaderni di studi semiotici, 12.
Petitot, J. et al. (1997). *Naturalizing phenomenology*. Stanford: Stanford University Press.
Piotrowski, D. (1997). *Dynamiques et structure en langue*. Paris: éditions du C.N.R.S.
Piotrowski, D. (2009). *Phénomènologie et objectivité linguistique*. Paris: Honoré-Champion.
Rastier, F. (2001). *Arts et sciences du texte*. Paris: PUF.
Rizzi, L. (1991). «Sulla conoscenza del linguaggio». *Cahiers Ferdinand de Saussure*, *45*301–316.
Salanskis, J.M. (2001). *Herméneutique et cognition*. Lille: Presses Universitaires du Septentrion.
Stewart, I. (1993). *Dio gioca a dadi ?*. Torino: Bollati Boringhieri.
Talmy, L. (2000). *Cognitive Semantics*, I-II, Stanford University Press.
Thom, R. (1970). «Les mathématiques modernes: une érreurs pédagogique et philosophique?». *L'age de la science*, *3*8–25.

Thom, R. (1972). *Stabilité structurelle et morphogenèse*. Paris: interÉditions.
Thom, R. (1979). *Logos Phénix* (pp. 377–388), in Critique, 7.
Thom, R, (1980). *Modèles mathématiques de la morphogenèse*. Paris: Bourois éditeur.
Thom, R. (1988). *Esquisse d'une sémiophysique*. Paris: interÉditions.
Thom, R. (1990). *Apologie du logos,*. Paris: Hachette.
Tomalin, M. (2006). *Linguistics and formal sciences. The origins of generative grammars*. Cambridge: Cambridge University Press.
Victorri, B. (1994). «The use of continuity in modeling semantic phenomena». In C. Fuchs & B. Victorri (Eds.) *Continuity in linguistic semantics* (pp. 241–251). Amsterdam: Benjamins.
Victorri, B. (1997). «Modèles en linguistique». *Encyclopaedia Universalis*, Paris.
Victorri, B. (2000). «Théories linguistiques et cognitition». *Incognito*, *16*(2)1–6.
Victorri, B. (2004). «Les grammaires cognitives». In C. Fuchs (Ed.) *La linguistique cognitive*. Paris: Ophrys.
Victorri, B. (2005). «Les mystères de l'emergence du langage». In J. -M. Hombert (Ed.) *Aux origines des langues et du langage* (pp. 212–231). Paris: Fayard.
Victorri, B. (2007). «Termes en kw-, récursivité et origine du langage». In M. Charolles, N. Fournier, Catherine Fuchs, & F. Lefeuvre (Eds.) *Parcours de la phrase,* (pp. 259–273). Paris: Ophrys.
Victorri, B., & Fuchs, C. (1994). *La polysémie. Construction dynamique du sens*. Paris: Hermés.
Visetti, Y. M. (2004). «Le continu en sémantique : une question de formes». *Cahiers de praxématique*, *42*39–73.
Visetti, Y. M., & Piotrowski D. (2014). «Connaissance sémiotique et mathématisation». In F. Lo Piparo, F. La Mantia & C. Paolucci (2014) *Semiotica e matematiche* (pp. 141–170) Versus, Quaderni di studi semiotici, 12.
Wildgen, W. (2009). «La rivalité historique entre une modélisation statique et dynamique des faits linguistiques». *in Histoire, Epistémologie, Langage*, *31*(1)99–126.
Zwart, J. W. (2011). «Recursion: a Layered-Derivational Approach». *Biolinguistics*, *5*(1–2)43–56.

Part I
Complexity and Linguistic Theory: Epistemological Questions

Chapter 1
The Game of Complexity and Linguistic Theorization

David Piotrowski and Y.-M. Visetti

1.1 Introduction: An Age-Old Debate

The matter at hand being language and, more broadly, given that we are dealing with scientific issues and even common affairs, and since these are addressed from the standpoint of knowledge or of practical success, the notions of simplicity and of complexity, the effort of distinguishing them and the tension they acquire are all elements which permeate the reflection presented here. We know that this translates into a great number of truisms, of commonplaces, of «multi-purpose» topoi that are as generic as they are opposed: Does intelligence reside with simplicity or should it rather be thought of only in terms of a sound trajectory over an uneven topography? Should it be envisioned in terms of exclusive postures or rather as complementary parts forming any architecture of knowledge, parts which it will necessarily be compelled to assemble? The example of mathematics in this regard is edifying: having long been valued because it pertained to structures that were simple, eternal, and ideal in character, it demonstrated during the classical age, with analytical geometry and then with infinitesimal calculus, an unsuspected generative power, capable of accounting for an open variety of formal configurations, and this on the basis of relatively simple principles and objects stated from the onset. This formal generativity rapidly took hold of an increasing variety of natural complexes in order to make them intelligible and, as was once believed, to provide them with perfect objective determination. Thus, began the mathematization of nature which was previously believed to be opaque and profuse, escaping any true intellection. The Newtonian (and then Laplacian) mechanism became, during two centuries, the

D. Piotrowski (✉) · Y.-M. Visetti
EHESS—CNRS, IMM—LiAS, 190 avenue de France, 75013 Paris, Poitiers, France
e-mail: david.piotrowski@ehess.fr

Y.-M. Visetti
e-mail: yves.visetti@ehess.fr

F. La Mantia et al. (eds.), *Language in Complexity*,
Lecture Notes in Morphogenesis, DOI 10.1007/978-3-319-29483-4_1

main figure of that which serves both as ontological conversion and as program for the advancement of sciences. To start with proper idealizations, to reduce phenomena to what they could be in pared-down experiential spaces—such was the new meaning of simplification opened up by the advancement of the sciences. Once engaged using the appropriate mathematics, it was possible, without exceeding a framework of rigorous laws, to attain increasing complexities, and to progressively close the gap between the available formal diversity and a natural diversity now subject to a synthesis by composition of laws governing physical systems.

However, this way of approaching the complexity of natural phenomena, through a certain physico-mathematical perspective, was not without clashing with the realities of the biological and of the psychological, to which this type of reductionist strategy could not fully apply. This is because these involved the intuition of a complexity which was of a radically different nature, directly unfolding through the spectacle of life, and beyond the grasp of such approaches: the organic being opposed here to the mechanical, and the movements of living organisms differing from those of geometric moving bodies. It appeared that the intuition involved could not be reduced to any formal or physical framework of presentation but that it rather solicited a particular sort of intelligence, on a level differing from that of a principle of separation between the sensible and the intelligible.[1]

This tension between simplicity and complexity, be it with respect to a first apprehension of phenomena, of their "nature" if you wish, or concerning the construction of knowledge regarding them, evidently found itself included within the scope of linguistic reflection, long before the development of the modern sciences of language.

For example, an echo of this is found in the ancient debate opposing the *Anomalist* school of thought (school of Pergamon) and the *Analogist* school (Aristotelian school of Alexandria) concerning languages and the scope of the grammatical fact (Douay and Pinto 1991). Of course, questions of mechanics or of analytical geometry will have been absent from this debate.[2] For the participants to this debate, it is a matter of determining whether languages develop and are to be understood as following regular principles, having the status of law (illustrated by the morphological paradigms of declension and conjugation) or whether the regularities one may find are but an effect of usages, commanded by simple tendencies to regular repetition as well as by shifts that are unpredictable both in their occurrence and in their subsequence. Conversely to the analogists, who call upon a framework of strict legality commanding the order of things and their intelligibility (thus rejecting particularisms and exceptions as insignificant), the anomalists uphold an «idea of nature where the stream of multiple determinations, which do not exclude a part of indetermination, produces [...] hapaxes, unique cases which stand out as

[1]Separation redistributed by Kant through the relation between intuition and understanding.

[2]But «Varro excellently reminds that it was by borrowing from mathematicians their *proportional relationship* (*analogon*, in Greek) that the Alexandrian grammarians were the first to be able to clearly establish, in the form of tables, the complex paradigms of Greek flexional morphology» [translation], op. cit. p. 8.

exceptions» [translation] (op. cit. p. 13.) Consequently, «rather than irritating oneself with exceptions, [the scholar] must accept the idea they reveal—sometimes—a reality where specific or even singular characterizations are legitimate» [translation]. Thus, the anomalists, by opposing the analogist theory of metaphor which originated with Aristotle, developed a theory regarding changes in meaning to which they gave the name of Tropes (*tropos*: turn or turning; *tropè*: sudden change of state, windshift) [14]. Douay and Pinto point out that *anomaly* does not signify here the negation of *nomos* (custom, law), but that of *homaios* (unchanging, even, regular, constant)—in other words, the anomalists reject the principle of a simple law determining and regulating a universe of homogenous objects, preferring a conception where the regularities and genericities are adjusted in usage and in the more or less tensive and assimilatory encounter of singular events, and from this stems the connection with the contemporary simplicity/complexity debate: not necessarily because analogists or anomalists identified with definite antagonistic positions in favor of a drastic reduction to simplicity or, conversely, to an unyielding defense of complexity, but because their debates provides an example of a dialectic between two ontological and gnoseological poles, a dialectic to which they are more or less committed and which is more or less complex (!).

As we can see, it is clear that the anomalists were not so blind as to reject any notion of regularity or of organization with respect to language or to any natural phenomenon. But their conception of such regularity is radically less univocal and determinant than for the analogists, for whom the systematicity and generativity of the rules can countenance no exception, to the point where such exceptions would elude any measure of intelligibility. For the anomalists, the perspective must be encompassing: things do not simply proceed from a complete and sovereign legislative apparatus, and any organization provides as much for singularity as for systematic reproduction according to recognized patterns. It is thus the very image of genericity which is inflected, to the very point where, far from opposing genericity, singularity becomes its generative figure, in terms of origin as well as its recurrent entry into a state of crisis. So we pass from a conception of genericity as a classifying schema, or as a type to be instantiated, to a conception of genericity as a mode of deployment engaging unstable forms which would concentrate its principle. The mode of intervention of genericity is disrupted: it is no longer an inflexible or univocal law, but rather and from the onset a heterogeneous montage and transitory harmonization, an anticipation fed by singular events that have been more or less reduced and assimilated. It is, then, a single and same framework which must encompass and comprehend the genesis of all forms, be they normal or monstrous—the examples of chimera and of animals thwarting established classifications are among those alluded to by the anomalists.[3]

[3]Regarding this too hastily reconstituted debate, one could think that it is a matter of a bipolarity of the regular/monstrous or regular/pathological type, but in truth, it is an indefinite variety of forms exceeding any typology which should be put forth here—and, in parallel, the idea of regularity should be made compatible with that of indetermination—but this is a much more contemporary reformulation.

The controversy between anomalists and analogists, despite being focused on linguistic issues, concerns both natural and cultural forms, and addresses their widely shared principles of intelligibility. A polarity of this kind could very well shed some light on the later history of ideas about languages, during the period between the eighteenth and twentieth centuries. May we not recognize, among all semiolinguistic disciplines, the most radical of our modern analogists among the proponents of a reduction of complexity to a composition of mechanical laws pertaining to sets of well-typed entities? The question then arises as to why, in the modern era, it is the currants that could be associated with the spirit of the analogist grammarian schools that appeared to be the custodians of scientificity, rather than the apparently less prevalent others that may be said to have anomalistic leanings, and which are steered toward literary studies, toward erudition—in short, toward the humanities.

1.2 The Nomothetic Approach: Reduction to Simplicity and Determinant Laws

In order to respond to this, it is necessary to take into consideration the unflagging influence of a certain episteme which largely encompasses scientific activities and which positions all disciplines within a progression scheme of which the ultimate stage would be the achievement of formalization, as imagined by means of a model allegedly taken from the mathematization of nature, such as in the physical sciences. It is to this type of episteme that the explanation of any regularity, with the reduction of phenomena to a core of systematicities, would have been entrusted. All idiographic research works were considered as nothing more than premises or as marginal.

This distribution of the values of elements of knowledge goes hand in hand with a certain articulation of simplicity and complexity in a context of scientific production where an essential role has been played by mathematical and later on by logical/computer-scientific images.[4] In such an episteme, the objectification process specific to scientific knowledge has long been thought of as a reduction governed by the indispensable objectives of parsimony, clarity, and simplicity—this being

[4]Let us emphasize in passing that the mathematical or logical/computer-scientific schemes in question were not themselves bearers of the epistemologies or of the ideologies that they may have contributed in founding. Even if, initially, the logico-mathematical schemes could appear as univocal epistemological emblems, nothing would prevent them from ulteriorly being requalified and put in the service of a plurality of epistemologies. Hence, the existence of computers and of programming languages by no means requires a logical and computational conception of the mind and of language. We may just as well consider this as an avatar of writing, having a form of hybridization with the idea of mechanism. Thus, *a contrario*, computer science has given rise to an unprecedented proliferation of diagrams and images which have contributed to the development of topological, dynamic and genetic conceptions, initially undervalued to the benefit of a purely computational vision.

applicable to the categorical apparatus and the associated laws, in particular combinatorial laws. This simplicity allows for the intelligibility and complete determination of the object, homologous to the objective of predictability advanced by the natural sciences: a simple yet powerfully generative explanatory core remains at the basis of the project.

We see that this is at the opposite of the recognition of a fundamental complexity which would require to reconsider these reductionist schemas, or, at least, would lead to a change in their epistemological status, conferring them but a hermeneutical and pragmatic modality among others, within a broader epistemological undertaking.

Consequently, from the moment when the language sciences adopted the epistemic schema of the "reduction to simplicity," complexity could only manifest as a symptom: through the difficulties that the attempts at theorization face when conducted within the framework of a *Hypothesis, Deduction, Refutation* type epistemology, that is, as regards linguistics, by adopting postulates such as (i) an autonomous grammaticality placed at the core of languages, (ii) the necessary horizon of a logicization, and (iii) the requirement for "Laplacian" determinism—all of these postulates mutually reinforcing one another in a context of computerization, a "Turingian" mechanism then taking the place of the Newtonian–Laplacian mechanism.

The conception of linguistic knowledge, following the dominant models developed in the natural sciences, as sets of determinant laws, has, in a way, repudiated the richness of phenomena by rejecting its essential primacy—always with the suspicion that giving it to much consideration would divert from the scientific ideal of nomothetic simplification. To associate it with a legendary icon, some have purported to say that Galileo had began by studying the fall of bodies dropped from the heights of the Tower of Pisa rather than by studying the fall of swirling leaves.

The required strategy was thus to aim to define a stable, determinable core that was sufficiently generative in order to account for a certain "level" of language production. But, in doing so, the perimeter of phenomena was closed off, particularly with the adoption of a formalist perspective. The centrality of syntax will have set the focus on questions of nesting or of "ascent" (in particular, of sub-trees), and on questions regarding the position of indicators of modality or of temporality along the syntactic trees, etc. In terms of semantics, which was thought to be separable, one version or another of the logic of predicates was adopted. The favored questions were namely those concerning the scope of quantifiers, of anaphoric references, and of ambiguity, whereas the question of the formation and of the identification of lexical formants was abusively reduced to that of the logical identity of lexical atoms, considered as necessary starting points for eventual processes of contextualization, the study of which was postponed to an ulterior stage. More generally, language activity being conceived as a sequencing of sentences, themselves stemming from a model of propositional logic strictly dependant upon a rigorous syntactic typing, we understand that the attention was for a long time captured by the issue, thought to be essential, of the accurate discrimination of constructions in terms of grammatical admissibility.

In this epistemological context, facts said to be singular (interpretive openness as well as irregularity of the forms encountered) may have been granted different statuses. Either that, being considered as irreducible singularities, they will be excluded from the scope of the explanation, or that, approached in terms of contextual perturbations (that are in a way analogous to friction in mechanics), the attempt may have been made to reintegrate them by means of an external multiplication of interacting levels of analysis (syntactical, semantic, pragmatic).[5] Or, otherwise, they may have been integrated within the central apparatus by means of an internal multiplication of categorical and functional subtypes—but with the eventual blockage caused by a profusion which infringes upon the principle of a simple generative systemic core. Harris' distributional analysis programs (1960), and later on Gross' lexicon-grammar construction programs (1975), and the conclusions that may have been drawn from them are, in this respect, exemplary.

In short, a narrowly rationalistic view of the scientific endeavor, only concerning itself with apparent regularities, was embodied in a certain (combinatorial or computational) idea of system, a formal counterpoint to the determinant closure of the object. If several versions have competed to establish these new systemic ontologies, the main idea to retain will be that of *structure*, its inspiration being more logical than mathematical, and with it being generally construed following the model of formal writing, governed by notions of logical syntax, of calculability (Chomsky) or, more modestly, of combinatorics (features, semes).[6] Thus was approached, during the second half of the twentieth century, that which was designated by the expression "language system." The simplicity and parsimony of the principles and elements established at the onset did not prevent from thinking, as the history of mathematics and of physics has shown, that there was sufficient generativity to address all complications compatible with the type of objectivization sought and with the theoretical devices established for that very purpose. Such an idea of system seemed to provide the semiolinguistic disciplines with the guarantee of a certain gnoseological model lent to the natural sciences, associating atomism, determinant laws and the quest for formalization, but this time, directly in some logical language and not through the mathematical idealities of infinity, of continuity, or of space, all of these being favored in the physical sciences.

[5]The irony of this being that in order to establish its own autonomy, each component of the global architecture (each subdiscipline) recommends its own criteria and a very specific level of objectivity. But then, the encounter between the different levels believed to be separable becomes problematic. The strategy of enunciative linguistics, for example, was to refuse such a separation: without renouncing the idea of a core of meaning, nor that of tiered layers, they are conceived differently, within an apparatus which anticipates beforehand the linguistic value of the enunciative and pragmatic activity.

[6]All being ingredients proposed at times by favoring relationships over elements, at other times by conferring these a first form of individuation which then supports relational connections specifying their identity.

1.3 Complexity, Replayed in the Models of Complex Systems

It is, however, apparent that this type of acknowledged complication (stemming from a generative power supported by simple elements) is fundamentally distinguishable from other types of complexity schemes, which we can see at work in the domains of the biological, of the psychological, of the social and, as far as our particular interests are concerned, of the symbolical.

At the same time, however, and sharing with physics a part of these schematizations (starting with differential equations and, more generally, by adhering to a continuistic and dynamicist framework), another conception of systems was sought, one which was attentive, above all, to living forms, to their genesis, to their functioning and to their interactions within an ecology. First favoring a sometimes "phenomenological" approach with respect to a singular living phenomenon and its significant forms (gestalt, autopoiesis…), a sometimes eco-systemic approach (populations, resources, territories), or even a social one (alliances, hierarchies, roles, values), these perspectives had the ambition, via the notion of organization, of understanding the inscription of the biological and of the social within the order of nature, thus broadening the physical order beyond what an elementaristic reductionism could grasp of it. Namely through the two cybernetics (cf. Dupuy 1994), with the themes of self-organization and of emergence, transversal problematics concerning systems spanning "from the physical to the political" were progressively constituted, these being crucially dependent upon a dynamicist framework with continuistic tendencies. Thus, another relationship to complexity, immediately positive, could be outlined in different fields of knowledge. And it is, mainly, through the process of *modelizing complex systems* that the problematic backdrop shared by these various attempts was progressively made explicit, that is, through a *modelizing* process, one which is not necessarily in keeping with a specific theoretical apparatus, but underlain by epistemological questions which we will now examine.

At the historical source of complex systems, we often find the idea of building a variety of dynamic systems into a network, namely those systems stemming from physicalistic elaborations and borrowing from various mathematical formalisms: deterministic or randomly parametered differential equations, stochastic differential equations, partial differential equations, discrete stochastic processes. But these concepts are no longer solicited only in view of a quantitative determination of phenomena; they also serve to build *qualitative* models of them, models which will be steered toward the restitution of observable morphologies and structures, and which are *functional*, meaning that they reproduce, when required following adaptation, regular modes of interaction coupled with certain viability criteria.

A series of advances recorded over the past decades will thus have lead physicists, mathematicians, biologists, computer scientists, model developers in the fields of cognitive and social sciences, to set the foundations of transversal objectification frameworks for their various disciplines, the core issues being questions of stability and instability, of invariance and variation, of regulation and viability: all questions that have generally remained foreign to the elementarist and formalist approaches and, conversely, that have, in the complex systems approach, been placed at the foundation of the ideas of system and of systematicity.

Among the main aspects of a systematicity thus reunderstood, we will mention, given that there is no model that can purport to regroup them all: detection at several spatial and temporal scales (generally two: micro- and macroscopic); topological, dynamic and statistical characterizations; reciprocal determinations at the local and global levels; multiple dynamics of unit formation (fusions and dissociations; coalitions and competitions; increases, decreases; births and deaths); coexistence of various dynamic "phases"; adaptation and regulation (active preservation of the internal and external domains of viability); structural derivation by coupling with a specific environment (circular causalities); historicity and irreversibility; collections of behaviors centered around instable dynamics constituting the functional core of systems. From this set, it is easy to conceive that there is no need to trace a distinction in principle between functioning and change. Likewise, the question of forms and organizations can now be posed without having to choose between aspects on the basis of their being individual or collective, internal or external, synchronic or diachronic. This is so even if it appears that, in most cases, the models have difficulty restituting the multiplicity and intertwined character of the "levels," "modes," or "phases" of the organization.

With this new way of organizing and of understanding dynamic systems, such as borrowed from the physico-mathematical tradition, the principle of a transversality between the fields of the natural and of the cultural is outlined, one which should now be addressed. Is it a question of the generalization of a concept of natural organization which would go so far as to encompass the semiotic as well as the social (the questions of signs and meaning, of symbolic forms, of institution)? Or is it, conversely, a question of renewing our understanding of our relationship to nature in an explicitly hermeneutical and pragmatic approach that delivers the true meaning of any physicalistic reduction? Or is it the onset of a new ontology encompassing the act, the subjects and objects of knowledge, and of which the modelizing practice would be the mirror and medium?

1.4 Emergence: Between Determinant Judgment and Reflective Judgment

In order to clarify and provide some form of response, examine the distinction between *determinant* judgment and *reflective* judgment, introduced by Kant in the third Critique in order to account for the difference between a judgment that carries knowledge and which operates by object determination, and a judgment serving as attachment to a *meaning* delivered through the intuition of a phenomenal form. Reflective judgment, of which it is necessary to understand the possibility as well as the claim to universality, has been recognized and analyzed by Kant as being of two sorts, one being the aesthetic judgment, the other pertaining to a formulation of the finality of organic forms in a natural system (architectonics). In short, we may say that for Kant, anything which pertains to the assignation of value (within generalized aesthetics) pertains to the reflective. This also applies to judgments pertaining to the functions and finalities of totalities, in which the parts are articulated not according to the mode of assembly, but according to that of a global and reciprocal differentiation.

Such a return to the Kantian distinction does not hold a doctrinal value for us, but it does serve as a resource enabling to register a contrast between (at least) two regimes of thought, at work in the semiolinguistic disciplines. We would simply like to emphasize that under the scope of reflective judgments, according to Kant, we find the modalities of assertion of a sort of "knowledge" which may be deemed suitable for these disciplines. First, what Kant says about the aesthetic judgment may naturally be extended to any operation pertaining to the recognition of value, as the encounter of an intelligible through a series of sensible valences. We may thus understand, for example, that linguistics thought as being inseparable from rhetorics (that is, from poetics and from stylistics) would necessarily relate to judgments which Kant would have qualified as reflective, that is, as disengaged from the model of knowledge achieved solely by means of such understanding as was exposed in the first *Critique*.[7] Second, it may be conceded that the Kantian model of nature, combining teleological and architectonic analyses, is an acceptable precursor of what the past century's systemic analyses have proposed, by attempting to define types (genera, species) on the basis of discerning mereological and functional relationships. These relationships indeed always refer, step by step, to a teleonomy and to architectures that are global (these being, it is true, more or less explicit). They thus converge with the Kantian vision of nature as exposed in the *Critique of the Power of Judgment*, even if it proceeds from what Kant calls thought (i.e., a well-directed use of reason) and not, as such, from what he calls *knowledge* (a designation he reserved for the determination of phenomena in intuition by understanding). This Kantian view of nature implements something of an early systemic approach—proceeding, however, in its apparatus, from the unity and spontaneity of an *I think*.

The issue of the subject, which emerges at this point, should also capture our attention. We know that in the context of a quite largely shared epistemology, the place of the subject of knowledge is not, or seldom, thematized. The observing and intervening subject is substituted with an observation apparatus (a viewpoint enhanced with a qualification and measurement grid); thus, the whole practice and the gain in intelligibility stem from the specification of methods or of protocols correlated with the structure of the established objectivity—for example, the

[7]And this even if, for us, in the contemporary context, it would be suitable for such linguistics to invoke yet a different type of knowledge, left uncovered by the system of the three Critiques, and consisting in the exposition of a reasoned distribution of competing interpretive points of view regarding a same value-form.

method for reducing problems to a twofold movement of analysis and then synthesis, correlated with an atomistic (more generally, elementarist) image of the field addressed.

How does this viewpoint stand as regards complex systems? It has often been said, within interested circles, that the systemic fact in and of itself had for matrix the mostly tacit *relation* that a collective of subjects bear with the world and where a systematicity comes to their attention, interrogating them. If a certain influence of the hermeneutical currents of phenomenology cannot be denied, it had little effect upon the principles of construction and upon the way models were conceived: these, in most part, were positioned within a physicalistic exteriority, with an objectivity deemed independent from any acting subject, whereas the subjects themselves had the tendency to be thought of as simple components of a vast complex system encompassing the subject and the world.

In what concerns the method and the desired gain in intelligibility, three dimensions are essentially involved: (i) an articulation into micro and macro levels, (ii) a certain form of genetic constructivism endowed with the task of bringing forth a variety of organizations from an initial state, by progressing along a parameter (said to be temporal) governing the evolution (or, more abstractly, the transformations) of a network of dynamic systems,[8] and (iii) the outline of a local/global dialectic, affecting both the organizing structures and values.

Of course, a coherent usage of these models should brush aside any intention of causing the appearance of signifying organizations by way of emergence, from ill-calibrated initial states, that is, states which are too organized or too unorganized for a variety of interesting macroscopic configurations to be declined from them (that is, configurations conjugating anticipated typical cases with less predictable variations). The gain in knowledge here is to be sought in the quality of the deployed possibilities, as well as in the delimitation of the parameters leading to the evolutions observed in the fields of reference.

The superposition of two levels of analysis (micro/determinant and macro/qualitative) is certainly a great asset with respect to the problem of complex systems. Therefore, the notion of system, which was conceived within a formalistic framework as a regulated combination of types representing a certain nomothetical ideal of knowledge, was considerably mitigated by the idea of an emergent typology of objects referring to a qualitative reading of categorical identities, now devoid of a determining scope. The gnoseological role of categoriality was thus reconsidered: it is no longer the crystalline basis of determinant procedures, but rather a typifying recourse to emergent organizations.

However, we are still far from conceiving and from achieving an authentic set of genetic issues in which even the components deemed elementary and their interaction schemas would be configured within emergent systems from a substrate

[8]This temporal parameter t is in fact far from referring to a clearly identifiable time. It is better conceived as a parameter controlling the system's variations in state, the progressive interaction of the organization factors retained in the modelization. It is also in correspondence with the time (endowed with an elementary increment value) of the algorithmic process which accomplishes it.

which would lend itself to individuations, or metabolic integrations, such micro-protagonists[9] being under the grasp of totalities already bearing meaning (reminding in part of the Kantian idea of architectonics expressing the force lines of a teleonomy). Now, in the case of complex systems, it appears that we remain confronted with a dilemma: either to begin with an amorphous initial state, or at least a random one with respect to the global finalities at stake; either to define the values or functions of certain elements from the microscopic levels, thus contravening to the objective of finding them by means of emergence (generally achieved in models by means of learning or adaptation strategies).

This was particularly the case for cognitive or semiolinguistic models: given the ease of modelization provided by a certain mixture of mechanicism and finitism, reliance has often been placed upon an elementaristic representation of part of the basic levels, constituted by units carrying from the start systemic values assimilated to the numeric value of their parameters. In the linguistic field, for example, one may think of the connectionist modelizations where the input and output levels directly encoded previously identified words or features. It was then a question of bringing about the emergence, in the context of a more or less supervised predictive or categorizing task, micro-values conveyed by other units and connections (said to be internal). An analysis of the data (for example, of the type of main component) enabled to reveal a geometric distribution of emergent representations, correlated to the relevant functioning of the system. What can be found there, summarily and in qualitative and emergent form, has to do with categorization and organization into constituents, definitively reproducing the categoriality of traditional grammars, but without endowing it with the determinant powers it has in formal approaches.[10]

Indeed, in the project, the ideal of complex systems particularly sought to reveal regularities in the very functioning of material dynamic systems, which would achieve something of the outline of a true organicity, with its ecological implications. The utopia was to even generate the organic fact as such. But it remains that the dialectics established between global and local values remained dependent upon a conception of formants and of their functions that was too unyielding and summary. The issue being one pertaining to semiolinguistic disciplines, the complex systems, as were sketched-out in the connectionist models (and which can be seen as sort of elementary building blocks) would have needed to refer to other views of language activity. Carried by other linguistic theories in better sync with the global holistic and ecological project, they may have gone beyond the option of a

---

[9]By decomposition and synthesis, it is a biophysical framework; by diffusion, transformation, institution, it is a socio-symbolical framework.

[10]We can thus evoke the proximity, once observed, between Californian cognitive linguistics and connectionist modelization. In a way, connectionist modelizations in the tradition of *PDP* (Rumelhart and McClelland 1986) will have reproduced the ambiguities and limits of the cognitive linguistics from which they drew inspiration. Giving preeminence to perceptual-type mechanisms, they conceived of them according to a mentalist and representationalist mode, and since it is a matter of language, they will not have succeeded in dissociating themselves from the traditional schemas of a sentence-based grammar.

representationalist and mentalist neuromimetism. But for that, it was necessary to assign the envisioned gain to its rightful place, and to recognize with greater clarity not an undertaking in pursuit of object determination, but a framework enabling the search for original regimes relating to formation and to the assignment of values, thus breaking with logico-grammatical systematicity.

In truth, the first contribution of these models was mainly to have illustrated the interest and fecundity of the principles said to be emergentist. It is only by effect of a once prevalent illusion that it was possible to believe that with them began a movement of *determination* with respect to semiotic phenomena and events of signification. The issue here is indeed the question of the organizations in that they are not simply formal and/or material, but that they metabolize and circulate *values*.[11] It is important to bear in mind that such values, as objectivated as they may seem in a model, proceed from another modality of existence, one which is both vital and semiotic, and which echoes the Kantian faculty of reflective judgment: a judgment in which the recognized meaning is asserted in a solidarily perceived form (without, however, such solidary grasp being reducible to object determination).

Approaches in terms of complex systems have nevertheless shown that it was of no use to restrict the operative field in order to determine independent objects that carry their own schemas of determination. The double micro/macro and local/global requirement was their fundamental principle. But beyond the elementarism (criticized above) persisting in the microscopic basis of models, they have misunderstood the fundamentally interpretative nature of the values, which is manifest in a particular difficulty in analyzing the reciprocal conditioning of the local and the global. There is, in that, kind of a necessary opacity, which is characteristic of the involved hermeneutical dimension. We will see the mark of a tacit depth which cannot be resolved in an exhaustive analytical explicitation and which refers to a community of meaning distributed over various instances (institutions, actors, techniques…), sources of capabilities but also of normativity. This opaqueness should therefore be found in the inception of any knowledge, as a presence, in each of its steps, of the necessary backdrop of pre-comprehension and of the pursuit of participation in some kind of social interaction.

The typical epistemology of complex systems will have retained from this opacity, from this tacit profoundness, only a physicalistic image, where the global effects of a multitude of locally determinant schemas are usually analyzed (sometimes made less strict by the introduction of a probabilistic dimension). This complexity, in the technical (physico-mathematical) sense of the term, is then a surrogate for hermeneutical opacity, which we could claim is inherent to an appropriate scientific apprehension of the phenomena under consideration. Thus, the unfettered undertakings engaged in during modelization (choice of architectures, of dynamic laws, of systems of features, of probabilistic inferences) steer away from a full clarification of the phenomena according to a determinant mode:

[11] The notion of value is understood here in the broad sense encompassing all issues at stake in the interactions, be they internal or external, which constitute and are created by the life of systems.

in the end, this comforts the aforementioned underlying reflective dimension of complex systems models. This also explains that applied to semiolinguistic data, such models have hardly contributed in revealing other categories than those stemming form grammatical and lexicological traditions, such categories being reinterpreted through the prism of a conservative reading.

1.5 The Middle Path of Complex Systems: Assessment for Semiolinguistics

This physico-mathematical take on a hermeneutical opacity, which we have said to form an integral part of any authentic approach to complexity, has nevertheless participated in a positive modification of the gnoseological landscape. There was a hope that it would produce a new comprehension of the role of modelization in the fields where it is a question of configuring domains and forms with respect to cultural and social values. In many cases, and if we want to go beyond a first promotional level of exposition, we clearly see that the proposed models have no pretension to causality or determinacy (their extreme simplicity attesting to this), but can rather be identified to exploratory practices, to a search for intelligibility having recourse to a kind of dynamic diagrammatization, a physicalistic sort of "rigorous fiction," at least one which is partially controlled, or even to another sort of experimentation by means of computerized devices (cf. Piotrowski and Visetti 2014).

It is therefore necessary to recognize that the issues of complexity addressed within these frameworks favor another view with respect to modelizing reason. It is thus not so much a question of predicting figures as of deploying them, be they typical or exceptional. Here, the images and levels of organization compose an always open repertoire.[12] An essential part of categoriality can be reunderstood as emergence, and the results resemble qualitative portraits rather than quantified stipulations of the phenomena under study. Hence the meaning of numerous structures and parameters remains suspended, being suggestions more than determinations.

This, and even when the configurations examined stem evidently from a physical order, we observe that in a majority of cases, the scope of the results can never be fully assigned in function of an exterior and anonymous objectivity. It must always be guaranteed by human actors, alone capable of accounting for them and of connecting them within devices of knowledge: this situation differs profoundly from those which can be observed in other areas of the natural sciences, where well-characterized determinant causalities can effectively participate in encompassing montages, like machines. We can imagine *a fortiori* what can be of this

[12]As evoked by the title of the pioneering work by H. Atlan *Between crystal and smoke* [*Entre le cristal et la fumée*, 1979].

when the modelized values claim to represent cultural values. In such cases, times, places and materials are but indications for interpretive work; they should be considered as vehicles that are transitorily endowed with values and not as components which define them as such.

For certain, the issue of complex systems appears to have indeed traced a sort of middle path for the recognition of complexity and for the scientific and technical attitude it should elicit. Be it out of necessity or out of conviction, a form of elementarism was maintained, for example by preassigning the functional values or significations of certain microscopic components in the models. While organization is said to continue at an emergent macroscopic level, the initial states were for the most part conceived from purely local viewpoints (microscopic data and schemas), without accounting for other organizational characters which involve a more global order (macroscopic). The result has been a frequent pursuit of the chimerical project of understanding the apparition of organization based on a poor qualification of the initial states, which is reminiscent of the tabula rasa of radical empiricisms.

Indeed, a new figure has thus been proposed for scientific knowledge, reflecting an enhanced image of nature, inspired by certain fields of the physical sciences, and promoting a fundamentally dynamic and statistic idea of organizations.[13] By their application to new fields, these physics-inspired models have ushered different interpretive processes in that they comprise a certain part of opacity with a weakly defined gnoseological status (but not devoid of hermeneutical reasons, as discussed above). But a certain scientistic perspective, which dominates in the fields in question, has impeded a full acknowledgement of the reflective rather than determinant scope of most models. Their most authentic value proceeds indeed from the opening of new regimes of intelligibility, which rest upon the unencumbered exploitation of dynamic schemas of understanding, in close relationship with an iconic and diagrammatic level of understanding.

Such a conception of complexity, overtly naturalizing and dynamicistic, and which would have aimed to be authentically genetic, implicitly and unknowingly returns to a certain number of *themata* proceeding from the hermeneutical currents of phenomenology. It may have been possible to think that this would favor the development of problematics pertaining to cultural and semiotic facts which acknowledge the complexity of their material, and which seek to understand them otherwise than by means of a “reduction to simplicity.” In keeping with this hope, and as we have already emphasized, continuism and dynamicism made it possible to open up a new horizon of intelligibility, particularly with respect to linguistic facts. Exposing what is discrete and elementary as effects of discontinuity within a continuous substrate, it was then possible to reverse the principle of composition by the assembly of pre-given parts in order to substitute it with that of differentiation within a general plan providing for a double circulation between wholes and parts,

[13]It is these schemas of understanding and figuration (in torrents, populations, forms, metamorphoses) which then occupy the center of the reconstructions, instead of the ideas of writing and of rewriting which represent another vision of temporality, in thought and in history (theme noted by the philosopher C. Malabou).

between overall and constituent morphologies, in order to finally break with the idea of a simple categorical and combinatorial foundation. Under this new light, the issue of structures was put into relation with the very broad question of forms, itself conceived within an emergentist framework. Thus emerged the possibility for a dynamicist reconstruction of all structural problematics with respect to the signifier as well as to the signified.

Moreover, another vision of regularities appeared: being multiple and non imperative in and of themselves, constituting more of an emergent horizon than an initial foundation, it was proposed that they be understood as proceeding from usages which, by way of contrast, would cause the appearance of hapaxes. Any identity, be it lexical or categorical, constitutes but an effect of repetition induced by a series of occurrences, and this makes of the hapax not an occurrence to isolate but a natural event, a potentially promising vagary inductive of new lineages. In short, regularity appears not as a determining functioning principle, but as a motivating and reflecting horizon, and the complex systems model types could help to specify this portrait. This connects with the debate (cf. supra, Introduction) between the analogist and anomalist schools, this time legitimizing the anomalist viewpoint—the analogist position being but a limit case induced by a certain systemizing view or, even, by a normative program.

1.6 What Do the Facts Tell Us? a Few Remarks

1. *Evolutionist Models*. Let's first mention the models which attempt to clarify such or such stage in the phylogenesis of language: models of emergence pertaining to syntax and to regulated combinatorics; models for the formation of a shared vocabulary among a collective of agents endowed with rudimentary perceptual capabilities (Dessalles 2000, and many works buy L. Steels and col.). Without going into detail, we will simply identify the strongly elementaristic character of such modelization, with discretized primitives in terms of sound and meaning given at the onset. There is nothing here, in these particular cases, that needs to be added to the considerations that we have formulated above, be they positive or negative.
2. *Connectionist Modelizations in the Tradition of PDP*. Although these are not, as such, complex systems but rather elementary building blocks which may have been integrated within broader architectures, these models nevertheless still embody a great number of the general characteristics examined above. The emergent categories or structures are distributed over collectives of units serving as a macroscopic level. Qualitative types of regularities or of systematicities manifest within, at the same time that via learning algorithms, a series of micro-features establish themselves, globally conditioned by a task of the prediction/categorization type over input/output layers.

   On a linguistic level, and despite a relativization of the importance of syntax, the debate has remained dominated by the question of knowing what, within this framework, became on the one hand of constituency and, on the other hand, what became of compositionality, conceived more broadly than in the "classical" paradigm as an always contextual synthesis of pre-identified elementary contributions (in the form, for example, of dictionary entries). As we have said, the categorical aspect retrieved by such means hardly distinguishes itself from the commonly shared one of "parts of speech." For want of turning towards linguistics bearing a global theory of interpretation and of the semiolinguistic field, it was impossible to relinquish a certain form of elementarism. For instance, "words" could not be individuated—assuming such a thing would still be necessary—within the framework of a global figure under construction, since the models took them for starting points. And if there were formants which were not aligned with the syntactic constituents, or which did not intuitively comply with a formal combinatorics of the R(A, B,...) type, it became difficult to evoke them.

From the standpoint of theoretical references, these models have found the path of a connivance with the mentalism and schematism of cognitive linguistics. They have also led to somewhat different research endeavors which approached the formation of utterances starting with the optimization of a weighted sum of constraints (Smolensky and Legendre 2006). In the latter case, all the necessary syntactic-semantic features are given at the onset, and the modelization effort pertains to learning about the weighting of constraints. And, contrarily to other connectionist works which have maintained a rather vague relationship with linguistic theory, Smolensky's modelization efforts readily present themselves as the applied facet of an original theoretical research, in the form of harmonic grammar and phonology.

But regardless of the status of all of these models with respect to the linguistic disciplines, and whether the authors endorse a physicalistic interpretation of them or not, these works remain attached to the idea of a full identification of any quantity by spatialized individuals. Thus, localization and numerical evaluation are made into the core issues, relegating interpretation to the backdrop as a trajectory and as the assignment of meaning—which should never be but an element of support for the pursuit of such a trajectory.

3. *Textual and Lexical Statistics: The Shadow of Complexity.* One of the ambitions of these complex models will have been to come into contact with very broad sets of data constituting an ecologically plausible portrait of empirical fields. Without necessarily referring to highly specific linguistic conceptions, another scientific approach was then favored, based on the confrontation with consequent masses of data, through the mediation of statistical and digital instruments. Participating in this movement, some research works can be evoked here, which, without adhering to modelization principles and objectives specific to complex systems (themes of emergence, double micro/macro level...), have thoroughly exploited the same statistical prism (in the extended sense of enumeration or identification of co-occurrences) and, from there, have resulted in original linguistic and textual characterizations or conceptualizations. Beyond

the classical analyses aiming to characterize texts and authors (tradition of textual statistics), we can mention new conceptions of the lexicon: the notion of *lexical item* (J. Sinclair); or of the reconstruction of the lexical value as the distribution of a potential over a semantic space constructed from sets of synonyms (Ploux and Victorri 1998).

The linguistic entities thus delineated comprise a form of indetermination, induced by the statistical instrument. Given such a fecund material, one may be tempted to see there a new kind of idiographic conception of linguistic knowledge, where regularities are not separable from the data pertaining to a specific corpus. We do observe, however, some reminiscences of a more directly nomothetic ideal, in which the authors are tempted to erect such regularities into laws. Thus processed, these corpora find themselves to produce the image considered to be the most generic of linguistic manifestations, and the semiology they induce fully replaces any phenomenology of language activity, as a necessary counterpart to linguistic knowledge.

1.7 Concerning a Few Linguistic Theories

Before concluding, let us once again set our view on complexity, but this time in order to succinctly qualify the way in which it connects with a number of linguistic theories and establishes gnoseological, epistemological and formal perspectives specific to each. This will be done without any particular reference to the universe of modelization and to the way in which it approaches and presents complexity. The following remarks are only provided as suggestions, without any claim to a demonstrative value. They concern a few important linguistic theories that are suitable to be confronted with the theme of complexity. Those which we have retained share the characteristic of placing the issue of meaning at the core of their apparatus.

Hjelmslev. Glossematics presents some features of a nomothetic ideal doubly marked by the relational doctrine of language following Saussure, revisited by various structuralist currents as well as by a whole logician and algebraist approach, under the acknowledged influence of Carnap. It performs a reduction of complexity to simplicity that does not refer to a free and unlimited generative process, but which operates on the basis of a limited class of fundamental concepts (a finite repertory of descriptive concepts). To this conceptual architecture is correlated a method of application of such concepts to empirical data, which are texts. These texts, however, are not the objects of the inquiry as such, but simply the physical environment where sentence-related configurations are found. A certain relationship to linguistic phenomenality is thus established, on the one hand, through the choices of the initial textual set and, on the other hand, in the adoption of fundamental categories pertaining to the effective presence of the relational fact within the material, which thus acquires the status of "language substance". However, and we must insist, the text here is but an environment for the acquisition of semantic and phonic elements, and not an object for which a global justification is sought (be it rhetorical, argumentative, stylistic…). Although it is very rigorous in its expectations, this theoretical apparatus does not exactly meet the nomothetic ideal, since it is an issue of first constituting an apparatus and a method capable of accounting for, that is, of "describing and explaining," in a single movement, the extraordinary diversity of linguistic forms. It is therefore not a question of asserting generative and determinant laws for each language, and here, algebra has more of a notational than calculatory status. The languages reconstructed following this mode are indeed systems, not in the sense of a productive formal mechanics, but rather in the sense of a relational structure of which the dependency relationships with respect to the syntagmatic and paradigmatic levels constitute the basic framework.

Cognitive Linguistics. As we know, cognitive linguistics—or it may be better to rather speak of cognitive grammars, which is not trivial in what concerns the issue of complexity—have shown how to connect perceptive constructions and meaning constructions without necessarily confounding them. These schools are thus relatable to the thesis of a certain linguistic schematism of a grammatical nature, which organizes any effect of meaning. In order to grasp this grammatical core, an author such as Talmy (2000) will refer to the notion of imagery (scene structure and division between actants, perspectives and trajectories, distribution of attention, system of forces). Langacker (1987), for his part, has conceived of remarkable diagrammatics in terms of fidelity and systematicity regarding his theoretical construction, diagrammatics to which it owes part of its success as well as its delimitation within a summarily spatialist problematic. Langacker's diagrams provoke a double resonance in the linguist's imagination: on the one hand, they resound in the recourse to a simplified mereology of the visual field as it would be configured by the utterance, and, on the other hand, they resound in the parallel application of these same "perceptivistic" schemas to the semiotic structure of the utterance (enabling, namely, to reinterpret the organization as constituents). These devices enable a corrosive criticism of the autonomy of syntax and of strict compositionality in semantics, at the same time as they enable to reject the logical formality and generativity which are usually at their foundation. In their stead, we now find semantic grammars from the onset, which provide the configurational framework for all effects of meaning.

These few reminders already suggest the way in which cognitive linguistics both acknowledge and simultaneously avoid the issue of complexity. In keeping with the necessity of addressing complexity, we will note the following points: (1) languages and language activity are no longer associated with a purely immanent functioning, but, to the contrary, are deployed through a permanent coupling with other cognitive functions, these being mostly perceptual (mainly visual), (2) the schematism which forms the explanatory core of linguistics is in close genetic relationship with the processual routines we believe to be characteristic of mental and cerebral functioning (of which connectionist networks would offer the most appropriate picture), (3) concerning the objectives assigned to linguistics, no longer is there a search for a generative core as such, but simply for a descriptive efficacy which rests upon the principle of a sort of gestaltist compositionality that can be identified in the analyzed syntagms, (4) following this, the empirical contact by means of which the theory adjusts itself is less a question of refutation than it is a certain form of hermeneutical attention supported by diagrammatic imagination.

But these openings in view of the recognition of the intrinsic complexity of the object of linguistics are promptly reduced in their scope: (i) a clear separation is introduced between the configurational structural dimension of meaning (the grammatical stratum which traverses all units) and the notional or conceptual dimension of meaning (all the rest), (ii) by such means a certain form of homogenizing simplicity is thus restored for this diminished object. This correlates with the pursuit of a form of theorization purporting to be determinant and which may have lead some to believe that the schema of scientificity by "hypothesis and refutation" was still operative here. More so that the explanatory principle consisting in understanding the plurality of acceptations by derivation from highly unifying schemas (or from a reduced collection of prototypes) almost marks the return, in terms of grammatical semantics, to an issue of generativity, despite it having been initially brushed aside. The invitation to open up the semantic process to not only its perceptual dimensions, but also to its encyclopedic and pragmatic dimensions points in a positive way toward a polysystematicity echoing the theme of complexity. But this perspective finds itself to be disappointed by the subjection of the inquiry to a core which is postulated as homogenous and coherent. To this is added the eminently disputable build of the schematic model retained for organizing the grammatical level and for simultaneously associating it with a visual perception of space (Cadiot and Visetti 2001; Visetti 2004a, c). Moreover, the emergentism put forward [Langacker or Talmy, then revisited by modelizers such as Regier (1996), Petitot and Doursat (2011)] remains very superficial, inasmuch as the structure revealed in the utterances proceeds from patchings between highly preconstituted units (words), beyond any logic of emergence and from underlying microscopic processes.

Thus, three major shortcomings impede the diagrammatic facet of cognitive linguistics. On the one hand, given the absence of a veritable dynamicist problematic, any *semiogenesis*, with its own (de-)strabilization dynamics and its open variety of emergent formants (expressions as contents) is ignored. On the other hand, the action which is speech is seen only through the action of schemas that concern only a small part of the subjects involved. Finally, the textual and dialogical dimensions are omitted or reduced to thematic formations on which, in any case, the diagrams have no hold: this is one of the major shortcomings of utterance-centered linguistics, which pursue the objective of characterizing at this stage a level which is consequent with respect to the forms without accounting for encompassing genres, with the dialogical, topical, and dialectical structures organizing discourse.

Enunciative Linguistics: Brief Remarks. The preceding criticism, indistinctively directed toward all utterance-centered linguistics, would deserve to be further developed on a case by case basis. For example, Culioli's *Theory of Enunciative Operations* (1990) is acknowledged as having directly linked linguistic analysis to the opening and organization of an enunciative scene proceeding from a set of acts. In parallel, the receivability criterion, usually addressed at a grammatical level, was enriched by new regimes of good formulation, indexed with respect to enunciative instances and situations. The idea of system took a backseat, whereas the idea of generativity was entirely reconverted in the perspective of a sort of genetic schematism attached to each singular unit. This approach did indeed recognize certain dimensions of linguistic complexity, but at the same time, there was a belief that it could be confined: on the one hand, by associating the exercise of language with a series of acts subsisting by themselves, since they proceed from generic types of which the identity can be understood by virtue of an isolation within the scene of enunciation; on the other hand, by positioning the level of "schematic forms" as a proper object of linguistics, in opposition with the abundant notional complex opened up by the lexicon and which is then thought of as a substance offered to the modeling work of schematic forms.

Interpretive Semantics (F. Rastier). In contrast with the aforementioned linguistics, Rastier's Interpretive Linguistics (1987, 1997) is the theoretical translation of an epistemology which recognizes the text as the true object of knowledge to be constructed in this regard. This text is not solely considered as the manifestation of abstract, narrative, argumentative, or other kinds of structures, which would be the only ones of importance. It must first be considered as a linguistic composition, as a field of expressive forms (inextricably expression and content), of which, within a reflective undertaking, the organization must be elucidated, namely on the basis of a network of genres and of other texts. Already, seen like this, the complexity of the object becomes apparent. The variational facts are primary, and the typologies to which they will be related provide them with a framework without, however, determining them. Furthermore, the text is thought as the medium for interpretations adjoining trajectories, themselves dependent upon points of view, reading projects, and conditional norms.

This is to say that the text is at the crossroads of several systemic branches of which the modal force, eminently variable, prevents any clear separation between system and norm, and likewise prevents any synthesis of this multiplicity into a single encompassing system. This marks a distinction with respect to both a principle of immanence and the idea of nomothetic knowledge. The milestones and results of the interpretative trajectories are then represented as distributions and configurations of *semes*. The seme, following a lineage tracing back to Saussure and including Hjelmslev, is conceived as a differential entity bearing the characterization of a relationship between two semiotic formations drawn from the backdrop of a class or set. All these elements participate in revealing complexity as an essential characteristic of the linguistic object, and analyses conducted in such a manner, far from impoverishing the text, reveal an original texture within, against which are highlighted veritably linguistic formations constituting that which can rightly be called the "text's textuality."

However, these methods are not without avoiding in some way a few fundamental aspects of complexity, namely the dynamicist dimension, which we would expect to manifest here in terms of (de-)stabilizations, of transformations, of formal metamorphoses within fields. In this perspective, it would be suitable, for example, to review the notion of difference, constitutive of that of seme, within an explicitly dynamicist (microgenetic) framework, where it could then become a *differentiation* and be considered in its various phases, thus enabling to relativize the discretization imposed by the semic analysis grid. Likewise, it would be suitable to restitute the notion of *transposability*. Indeed, if we were to limit ourselves to relying on the conceptions inherited from structuralism, we would not understand how it could be founded that a feature or a seme, constitutively dependent upon a determined class of interdefinition, could appertain to a class or transpose itself from one class to another, or even belong to no class in particular. Now, it seems essential to make intelligible this possibility, without loosing differentiality. It is, finally, the notion of *transformation*, also inherited from structuralism and serving as a token of the systematicity at play in the comparativist approaches, which should be further reexamined.[14]

Now, despite the initiation of a rewriting of *Interpretive Semantics* in continuistic and dynamicist terms, it clearly appears that these tensions have not been resolved. The concept of seme is now stretched between several definitions: the extremity of a binary relationship between sememes (1987), the smallest unit of meaning defined by analysis (2001), and, more recently, an interpretive moment, or simple feature participating in the description of semantic forms.[15] The concepts of inheritance, of afference, of inhibition, of virtualization, and of the actualization of semes, central in *Interpretive Semantics*, have hardly been reexamined in light of recent evolutions. The notion of transformation (from one form into another) will have lead to the individuation of semantic states in exclusive terms of figure/ground structures (rendered by semic arrangements), in order to attribute them as such to attested forms, taken as sole benchmarks for determination and repetition. The structuralist backdrop thus remains present in these movements of comparison/transformation, which sometimes appear to present themselves as the transcription of genetics as such (which, in the case of diachronic vocabulary studies, almost ends up assimilating itself with the historical dictionary genre, as in Rastier and Valette 2009).

[14]Cf. *a contrario* the propositions developed in (Cadiot and Visetti 2001, Visetti and Cadiot 2006) where it was a question, through a discussion with various linguistic theories including *Interpretive Semantics*, of rethinking the semiolinguistic field in dynamicist terms, as a deployment of forms participating in several regimes or phases of meaning, each referring to a certain mode of linguistic genericity. We then call "motifs" the semantic formations which carry a genericity, said to be figural, that is characteristic of instable phases, and possibly present under various textual formants (e.g. morphemes, lexemes, idiomatic expressions, isotopies, etc.).

[15]Thus, the term "seme" does not appear in the preface to the 3rd edition of the theory's founding work.

We can thus regard the concept of seme as the nodal point of a sort of conflict between two epistemological orientations. In the first of these, an elementarist point of view and the methodical impression it produces are not relinquished, and semic analysis seems to refer to a set of elements crystallized by virtue of a coherent and encompassing notation, even enabling to reconstruct types from which the occurrences are supposed to inherit by default. In the second orientation, which is fundamentally freer, the seme is but a moment we believe to be fugitively stabilized in an interpretative trajectory—but then, how do we understand its transposability to other classes and other trajectories? How can we even understand the possibility of an identity maintained through time?[16] Rather than the consignment of a "stabilized" moment, it seems more accurate to see here the marking of an articulation and interpretive rebound point, the notational and diagrammatical explicitation of a differential arrangement, of which the semantic destiny is to be repeated being reopened each time. Now, in the second epistemological orientation just mentioned, there is something of an enduring belief in the existence of a well-delimited linguistic level, devoid of genetic depth, where sets of features of a purely diacritical nature would reside, although it is deemed to be the sign of a certain objectivity. In a way, the theory has not entirely relinquished, if not its objectivist claims, at least a sort of operational mastery with respect to meaning, these being characteristic of a past era of research. So there is some kind of recoil when faced with the effects of the complexity of language phenomena, affecting the accurate vision of the theoretical undertaking, from the moment it would proceed from an art of intervention and explicitation, rather than from an act of determination.

1.8 Conclusion

To conclude these few considerations, may we allow the authors to share what has served as backdrop for their work and that which has nurtured their reflection at the crossroads between the view on complexity discussed herein and the issues pertaining to linguistics. To put it shortly, we can begin by mentioning the general dynamicist conception which today constitutes the best scientific locus for a philosophy of becoming, such as evoked by the few following keywords: flux, spatiality, temporality, instability, stabilization, phase, emergence, graduality,

[16]Thus, the first of these epistemological orientations seems to allow for the designations or glosses of semes to have only a conventional status (albeit motivated since it deals with *lexicalization*), whereas the semes are conceived as "method invariants", which their name may indicate but does not constitute. In the second orientation, on the other hand, it is not so directly a question of "method", and we would expect the more opportunistic nature of the seme, which is then to be understood as an event within a trajectory (an "interpretive moment"), to result in a particular concern for its designation. But such is not the case. And when it does appear to be so, it is simply that the analytical notion of semic relation has been substituted with the more simply empirical notion of co-occurrence.

differentiation, coupling, continuous/discontinuous dialectic… Scientific imagination of this kind, fuelled by mathematics, has taken the form of modelizations with various rationalities; sometimes in view of a confrontation with the empirical, sometimes as a simple diagram with a reflective vocation, an explicit form of the principle of intelligibility.

Regarding mathematical schematizations and models, the capital event, as we know, was the opening with R. Thom's Catastrophe Theory (CT) of a line of work which enabled to reconstruct structuralism within a dynamicist framework and to move beyond some aporia related to its enclosure within a universe of discrete representations. One of the essential contributions to the theory will have been to substitute the logico-algebraic vision of genericity with the topologico-dynamic idea of an unstable rudiment, which anticipates relatively to the stabilization paths that help to articulate the central phenomena of value determination (shift, fusion, reciprocal differentiation, dialectic identification of opposites). Thus, the gap which had artificially established itself between issues of structure and of form could subside, the structural "formality" now proceeding from a general theory of forms, where it could be diversified, through a renewed concept of *morphology*. Forms, geneses, and even systems were now available as three perspectives for the apprehension of a same phenomenon. At the same time, a new path would open up to connect the intelligible with the sensible by more explicitly drawing the grasp of meaning and structures closer to the perception of forms. This brought us back to the phenomenological principle of a *primacy of perception*—understood as the generic modality of the access to anything which presents itself to us and which carries meaning.

Such was the prism following which we came across the issues of complexity, and namely those which have accompanied the development of complex systems, which at times appeared to be a quite fruitful development of the first universe opened up by catastrophe theory (explicitation of the double micro/macro level, systems architecture, adaptation and learning, great sets of data), and at other times also a regression (precocious discretization, elementarism, loss of the differential character of values which CT, on the other hand, had translated so well by means of an articulation between internal dynamics and external spaces).

With this simple reopening of the issues of structural concern—owing to its singular ingenuity—appeared the possibility as well as the desire to connect with a whole lineage of twentieth century works, foremost among which those of the great authors of phenomenology, in both its philosophical version (Husserl, Gurwitsch, Merleau-Ponty…) and scientific one (Berlin Gestalt, schools of microgenesis). From this point and steering back toward issues of language, we have undertaken, with a few others and in a manner somewhat removed from the reference works cited above,[17] to emphasize within the theoretical edifices the shared nature of perception, pragma, and language, the recognition of which appears capable of

---

[17]Including a distancing from CT, which however remains a foundational moment, for each one of us, in our progressions steered towards different horizons.

preserving that of the complexity specific to language phenomena. A complexity upon which we have striven to project the threefold perspective informed by (i) linguistic and textual theories of interest (among which a few have been evoked above), (ii) some phenomenologist authors of the Husserlian school and, finally, (iii) the dynamicist and continuistic approach. Our ambition was to grasp the phenomenon of language starting at its very root, by conceiving it, following a formula by Coseriu "not as an activity which employs preestablished signs, but as an activity which creates signs" (translation 1958, Chap. II, 2.2) as a *semiogenesis* operating in the field of speech.[18]

References

Cadiot, P., & Visetti, Y.-M. (2001). *Pour une théorie des formes sémantiques—motifs, profils, thèmes*. Paris: Presses Universitaires de France.

Coseriu, E. (1958/2007). *Synchronie, diachronie et histoire*, Traduction par T. Verjans de *Sincronia, diacronìa e historia. El problema del cambio linguistico* (1958). Texte français intégral disponible sur http://www.revue-texto.net/Parutions/Parutions.html.

Culioli, A. (1990). *Pour une linguistique de l'énonciation. Opérations et représentations*, T. 1. Paris: Ophrys.

Dessalles, J.-L. (2000). *Aux origines du langage. Une histoire naturelle de la parole*. Paris: Hermès-Science.

Douay, F., & Pinto, J.-J. (1991). Analogie/anomalie. *Persée, 53*(1), 7–16.

Dupuy, J.-P. (1994). *Aux origines des sciences cognitives*. Paris: La Découverte.

Gross, M. (1975). *Méthodes en syntaxe*, Hermann,

Harris, Z. (1960). *Structural linguistics*. Chicago: University Chicago Press.

Langacker, R. (1987). *Foundations of cognitive grammar*. Stanford University Press.

Petitot, J. (1992). *Physique du sens: de la théorie des singularités aux structures sémio-narratives*. Paris: Éditions du CNRS.

Petitot, J., & Doursat, R. (2011). *Cognitive morphodynamics: Dynamical morphological models for constituency in perception and syntax*. Peter Lang.

Piotrowski, D. (1997). *Dynamiques et Structures en Langue*. Éditions du CNRS.

Piotrowski, D. (2009). *Phénoménalité et Objectivité Linguistiques*. Champion: Collection Bibliothèque de Grammaire et de Linguistique.

Piotrowski, D. (2010). Morphodynamique du signe; I—L'architecture fonctionnelle. *Cahiers Ferdinand de Saussure, 63*, 185–203.

Piotrowski, D. (2011). Morphodynamique du signe; II—Retour sur quelques concepts saussuriens. *Cahiers Ferdinand de Saussure, 64*, 101–118.

Piotrowski, D. (2013). L'opposition sémiotique/sémantique comme articulation de la conscience verbale. *Versus, 117*, 27–52.

Piotrowski, D., & Visetti, Y.-M. (2014). Connaissance sémiotique et mathématisation, sémiogenèse et explicitation. *Versus, 118*.

Ploux, S., & Victorri, B. (1998). Construction d'espaces sémantiques à l'aide de dictionnaires de synonymes. *Traitement automatique des langues, 39*(1), 161–182.

Rastier, F. (1987). *Sémantique interprétative*. Paris: PUF, coll. Formes sémiotiques.

Rastier, F. (1997). *Meaning and textuality*. University of Toronto Press.

[18]Also, cf. the works of the two authors cited in the bibliography.

Rastier, F., & Valette, M. (2009) De la polysémie à la néosémie. *La problématique du mot*, S. Mejri, éd., *Le français moderne*, *77*, 97–116.

Regier, T. (1996). *The human semantic potential: Spatial language and constrained connectionism*. Cambridge: MIT Press.

Rumelhart, D., & McClelland, J. (Eds.). (1986). *Parallel distributed processing: Explorations in the microstructure of cognition*. Cambridge: MIT Press.

Sinclair, J. (2004). *Trust the text: Language corpus and discourse*. London: Routledge.

Smolensky, P., & Legendre, G. (2006). *The harmonic mind: From neural computation to optimality-theoretic grammar* (Vols. 1 and 2). Cambridge, Mass: MIT Press.

Talmy, L. (2000). *Towards a cognitive semantics*. Cambridge: MIT Press.

Thom, R. (1972). *Stabilité structurelle et morphogenèse*. Paris: Édiscience.

Visetti, Y. -M. (2004a). Le Continu en sémantique—une question de Formes. *Cahiers de praxématique*, 42, numéro coordonné par D. Ablali and M. Valette sur *Le Continu : du son au sens,* pp. 39–74.

Visetti, Y. -M. (2004b). Constructivismes, émergences : une analyse sémantique et thématique. *Intellectica*, 2004/2, *39*, 229–259.

Visetti, Y.-M. (2004c). Language, space and theory of semantic forms. In A. Carsetti (Ed.), *Seeing, thinking and knowing—meaning and self-organization in vision and thought* (pp. 245–275). Dordrecht: Kluwer Academic Publishers.

Visetti, Y.-M., & Cadiot, P. (2006). *Motifs et proverbes—Essai de sémantique proverbiale*. Paris: Presses Universitaires de France.

Chapter 2
Continuity in the Interactions Between Linguistic Units

Gilles Col, Rossana De Angelis and Thierry Poibeau

2.1 Introduction

Linguistic tradition has produced descriptions that can be assimilated (at least for the most part) to discrete models: discrete models of categories, discrete grammar rules and discrete lists of word senses, among others. This view is now widely discussed and more and more linguists agree that this model is partial and, more importantly, cannot account for the whole complexity of human languages. For example, Manning (2003) clearly shows that a discrete model of categories is both too poor and too restrictive:

> Categorical linguistic theories claim too much. They place a hard categorical boundary of grammaticality where really there is a fuzzy edge, determined by many conflicting constraints and issues of conventionality versus human creativity.
>
> Categorical linguistic theories explain too little. They say nothing at all about the soft constraints that explain how people choose to say things (or how they choose to understand them).
>
> (Manning 2003: 297)

Similarly, dictionaries are interesting tools for providing references and definitions for humans but they are very unsatisfactory outside prototypical cases. It has been shown by a wide range of different authors that definitions overlap, and that different word senses (i.e. different dictionary entries) can correspond to a same occurrence in a corpus. The notion of graded word sense has been proposed

G. Col (✉) · T. Poibeau
Laboratoire LATTICE, CNRS & ENS & U. Sorbone nouvelle,
PSL Research University, 1 rue Maurice Arnoux, 92120 Montrouge, France
e-mail: gilles.col@univ-poitiers.fr

R. De Angelis
Laboratoire CeReS, Université de Limoges, Limoges, France

F. La Mantia et al. (eds.), *Language in Complexity*,
Lecture Notes in Morphogenesis, DOI 10.1007/978-3-319-29483-4_2

(Erk and McCarthy 2009), which is at the same time interesting and intriguing since it is not clear how it could afford the traditional model of dictionaries.

New linguistic models have thus been developed that address (at least to some extent) some of the limitations of traditional linguistic descriptions. In our opinion, two key features have contributed to the renewal of the domain: (i) the availability of very large corpora (and even the entire Web) that make it possible to access massive data and sets of examples and (ii) new computational tools that make it possible to observe linguistic data as a complex system, i.e. as a system in which the key feature is the interaction of units between themselves. These interactions are complex, multi-faceted and multi-layered so as to give birth to dynamical and moving landscapes of linguistic behaviour. This has been stressed by different authors, see for example Girault and Victorri (2009):

> Il faut donc changer de cadre de modélisation si l'on veut utiliser pleinement le nouvel observatoire que nous offrent les nouvelles technologies: les linguistiques de corpus se doivent d'être des théories s'appuyant sur les mathématiques du continu.
>
> (Girault and Victorri 2009: 153)

In computational linguistics, continuous models were marginally used compared to approaches based on dictionaries and rule that were dominating the period 1950–1980. With the rise of machine learning methods (especially distributional learning and "deep learning", see Socher et al. (2013) and the last section of the paper for a quick overview), continuous models tend to now be dominating. It seems that it is not the case in linguistics yet, despite various attempts since one or two decades.

Continuous models are highly challenging for linguistics: if interaction is at the heart of the linguistic process, it seems necessary to explain what is the source of the interaction, how it can be defined and what can be its dynamics. There is at least two apparently opposite views, which could account for linguistic interactions and more particularly for their dynamics: a compositional one (i.e. the meaning of a complex unit depends on its parts) and a gestaltist one (i.e. the meaning of a complex unit is more than the meaning of its parts). In fact, these views are not contradictory but should be combined. In what follows, we will see how we can reconcile them thanks to the notion of «instruction» and more specifically to the very content of these instructions.

The notion of instruction is basically considered here as a means to assess the notion of continuity since the content of an instruction is in fact a dual process of convocation and evocation. All linguistic units are considered as giving an instruction, thus taking part to the global meaning of the utterance and at the same enriching its own local meaning.

Some of these models, combining topology and dynamics, are presented in Sect. 2.2. We then detail in Sect. 2.3 the notion of instruction in a specific model (*gestalt compositionality*) and the last section enlarges upon continuity in language modelling. We conclude with a summary and some perspectives.

2.2 Topological and Dynamic Models in Linguistics: From "Autopoiesis" to the Notion of "Instruction"

2.2.1 Short Review of Some Models: Autopoietic, Gestaltist, Morphogenetic

Maturana and Varela (1980) propose a topological and dynamical model of living systems inspiring a topological and dynamical model of cultural systems. First distinguishing between *self-referred systems*—only referred to themselves, as living systems—and *allo-referred systems*—referred to a context—they propose an epistemological model in which cultural systems are conceived as texts (Maturana and Varela 1980: xiii). Speaking about «systems defined as unities through the basic circularity of their production of their components» (Maturana and Varela 1980: xiv), Maturana defines autopoiesis as "what takes place in the dynamics of the autonomy proper to living systems" (Maturana and Varela 1980: xvii). Developing this concept, «notions of purpose, function and goal are unnecessary and misleading» (Maturana and Varela 1980: xix). Then being an autopoietic system means (1) to be autonomous, to present itself as (2) an individuality and (3) a unity, (4) and do not have inputs or outputs.

The notion of autopoiesis is based on the fundamental notions of *unity*, *organization* and *structure*. *Unity* is «an entity distinct from a background, [which] characterizes both unity and background with the properties with which the operation endows them, and specify their separability. A unity thus specified is a simple unity that defines through its properties the space in which it exists and the phenomenal domain which it may generate in its interactions with other unities» (Maturana and Varela 1980: xix). Once identified a system as a unity, we can repeat this operation of distinction—the first perceptual and conceptual operation made by the observer—in a composite unity, going on with the identification of the components constituting the unity as a whole. «The relation between components that define a composite unity (system) as a composite unity of a particular kind, constitute its organization. In this definition of organization the components are viewed only in relation to their participation in the constitution of the unity (whole) that they integrate» (Maturana and Varela 1980: xix). This specific definition of *organization* calls back the hermeneutic conception of text considered as a whole. So, adopting the autopoietic model in linguistic analysis, we have to consider this hermeneutical principle. Finally, the notions of *unity* and *organization* are in relationship with the notion of *structure*, identifying the actual relations, which hold among the components (Maturana and Varela 1980: 77).

Focusing on the notion of structure, we can explain different topological and dynamical models stem from different epistemological paradigms. For instance, the *gestaltist structuralism* (De Angelis 2014) carries out from the encounter between *structuralism* and *gestalttheorie*.

Gestaltist structuralism is based on the assumption that any perception has to be considered referred to a biological organism and a perceptive organization, what explains the common agreement of *Structuralism* and *Gestalttheorie* on the concept of *structure* (Petitot-Cocorda 1985: 29–33). Perceptions are considered as *structures* (*Gestalten*), organic unities individualized in the spatial and temporal field of perception (Guillaume 1979: 23–27). Structures are organized and complex, resulting from a process of generation in which the constitutive elements cannot be considered independently from the whole in which they are included. One of the basic principles of *Gestaltpsychologie* is that things we observe in the environment are perceived regularly as stable (Köhler 2000: 191 sgg.). So the object of *gestaltist structuralism* is the way structures maintain stable or change, and how we can identify them in spite of their evolutions.

These ways are investigated in particular in a *morphogenetic* perspective. The *theory of catastrophes* proposed by Thom (1972, 1974) makes a synthesis between two concepts, *morfogenesis* and *structure*, and two different structuralisms, *biological* and *linguistic*. Thom shows the way structures ensue one another, first identifying this phenomenon as morphogenesis, then identifying the processes able to create or destroy structures (Thom 1974: 3). In fact, structures can occupy a certain space and last a certain time (Thom 1972: 3). The possibility to recognize a structure as being always the same depends on the possibility to recognize a plurality of aspects identifying the structure itself as a same object of analysis. Finally, the concept of morphogenesis supposes to identify a certain discontinuity interpreted as a change of structures itself.

The morphogenetic model applied to structures in general is then developed by Petitot-Cocorda in a more specific morphogenetic model applied to linguistic structures in particular. As Thom himself stresses introducing his work (Petitot-Cocorda 1985: 8), the approach to linguistic analysis proposed by Petitot-Cocorda has an important ontological consequence: we consider structures as a reification of connections, so we recognize to the terms connected a *positional value* and we reject a priori all phenomenological aspects of the terms composing the structure as a whole (Petitot-Cocorda 1985: 15). This model adopted in linguistic analysis is based on a *dynamical topology* (Petitot-Cocorda 1985: 8), concerning places and connections, based in turn on a *differential topology* (Thom 1974: 4), concerning specifically the differential connections of the terms. So, supposing that the *gestaltist structuralism* is based on a topological model, despite of a logical one, Petitot-Cocorda proposes a *morphogenesis of meaning*. This perspective is also based on an hermeneutical principle: as Thom says, *topology* is a mathematical discipline which lets to consider *the passage from global to local* (Thom 1972: 169). Adopting this point of view in semantics, we can identify *semantic structures* as stable, complex, locally determined, and connected to other semantic structures, all together constituting a whole, a global structure which is also stable, complex and coherent: the *text*.

According to the *morphogenesis of meaning*, each structure is stable and (auto)governed by the connections defining the *positional values* of the terms involved: because it is topologically determined, the value of a semantic entity is defined only by the place it occupies in the structure. So the *positional values*, their stability and change, can explain the organization, the stability and the transformation of the semantic structures. Finally, the notion of *positional value* supposes the notion of *differential value* proposed by Saussure[1]: we can identify linguistic entities only by the differential connections they entertain in language.

2.2.2 The Articulation Between Global and Local Dimensions

The relation between *global* and *local* dimensions in linguistic analysis is essentially drawn by the gestaltist structuralism. However, this relation is stressed also in general linguistic. As Victorri and Fuchs (1996) remind us, formerly we can read about the relation between global and local in Saussure's *Cours de linguistique générale*:

> Il ne suffit pas de considérer le rapport qui unit les diverses parties d'un syntagme entre elles (par exemple *court* et *tous* dans *contre tous*, *contre* et *maître* dans *contremaître*); il faut tenir compte aussi de celui qui relie le tout et ses parties (par exemple *contre tous* opposé d'une part à *contre*, de l'autre à *tous*, ou *contremaître* opposé à *contre* et à *maître*).
>
> (de Saussure [1916] 1922: 172)

Developing this point of view in a different dimension, that is to say the one of the sentence, also E. Benveniste speaks about this relation in his work *Problèmes de linguistique générale*:

> Une phrase constitue un tout, qui ne se réduit pas à la somme de ses parties; le sens inhérent à ce tout est réparti sur l'ensemble des contituants. Le mot est un constituant de la phrase, il en effectue la signification; mais il n'apparaît pas nécessairement dans la phrase avec le sens qu'il a comme unité autonome.
>
> (Benveniste 1966: 123–124)

Nevertheless, the relation between global and local dimensions becomes crucial to identify *semantic forms* in the approach named *interpretative semantics* as developed by Rastier. As we can read in his work *Sens et textualité*, understanding a linguistic sequence is an activity consisting in identifying *semantic forms* which are already understood or produced during the analysis of text (Rastier 1989: 9). The *interpretative semantics* has to be conceived as belonging to the rethorical-hermeneutical tradition (Rastier 1989, [1987] 2009). In fact, one of the fundamental principles in hermeneutics concerns the relation between global and local dimensions of texts: particular entities are determined by their relation to the text conceived as a whole which is in turn composed by them. Nevertheless, we can recognize the same principle in the *gestaltist structuralism*.

---

[1]«Il n'y a dans la langue ni signes [dans le sens de "signifiants"—N.d.A.], ni significations, mais des *différences* de signes et des *différences* de significations: lesquelles 1° n'existent les unes absolument que par les autres, (dans les deux sens), et sont donc inséparables et solidaires; mais 2° n'arrivent à se correspondre directement» (de Saussure 2002: 70). Cf. also CLG, Part II, Chap. III, IV».

The notion of *semantic form* evokes the one of *perceptive form*: in fact, Rastier considers language in general, expression and meaning in particular, as perceptive objects, so he proposes the notion of *semantic perception* (Rastier 2009: xiv). What he calls *semantic perception* depends on a hierarchy of three kinds of data: the *forms*, the *grounds*, the *backgrounds* from which grounds and forms emerge, that is to say the paradigms of other grounds and forms concurrent connecting the actual semantic perception to the corpus of previous linguistic experiences.

Furthermore, the relation between *semantic forms*, *grounds* and *backgrounds*, evokes the way in which the *Gestaltpsychologie* explains perceptive phenomena, in particular the ones of visual perception (Köhler 2000: 188 sgg.). When a visual *form* is perceived, the one emerging by subtraction is absorbed by the *background* and its own *form* becomes invisible. However, when the last one emerges from the *background*, the first one disappears. The *form* emerging on the top and the *form* staying in the *background* cannot be perceived at the same time. So, the existence of a visual form depends on the visual unite emerging from the background. In parallel, in semantics a *form* implies the existence of a semantic unit, which acquires its *form* after being isolated from the *background*. The *form* stays on a semantic *ground* on which differently extended semantic *units* stand on.

In the *interpretative semantics* the relation between *forms* and *background* depends on the differentiation of *semes* (the smallest units of meaning). The semes are distinguished as *generic* and *specific*: the first ones represent the *ground* of a lexical class—or an *isotopy*—from which the second ones emerge identifying the *semems* (the content of a morpheme) constituting this class as *semic molecules* (a group of *semes*, not necessarily lexicalized) which go through the text. So, we can realize in semantics what happens for the perceptive forms: during the interpretation of a text, semantic forms emerge from the background, distinguishing themselves from the other ones dispersed in the text.

So the sense is reproduced tracing a network of *semantic forms* laying on a network of *expressive forms* showing the *interpretative paths* of the text. The network of *semantic* and *expressive* forms is a global network constituted in turn by different local networks. As Rastier (2006) explains, concerning the dimension of meaning, networks are made by *semic molecules* and *semes* constitute their hubs; concerning the dimension of expression, networks are made by phemic molecules and their hubs are constituted by *phemes* (the smallest units of expression). The possibility to cross these two textual dimensions, meaning and expression, depends on the *interpretative paths* which take shape in the text and explain the bustle between a local and a global dimension according to an hermeneutical approach.[2]

[2]The simplest way to avoid answering the question is to consider the text as a sign. This solution is chosen by Pierce as well as Greimas and Eco (cf Eco 1988, p. 32: "le Message équivaut au Signe"). Yet, this solution pays little attention to the different levels of complexity characterizing signs and texts, and above all it prevents conceptualizing the impact of the global on the local, i.e. of the text on the signs (Rastier 1997).

> Le sens consiste pour l'essentiel en un réseau de relations entre signifiés au sein du texte—et dans cette perspective, les signifiants peuvent être considerés comme des interprétants qui permettent de construire certaines de ces relations. Elles demeurent de type perceptif: estimation de similarité, reconnaissance de forme, catégorisation.
>
> (Rastier 2001: 189–190)

According to *gestaltist structuralism* (Rastier 2009), the notion of *semantic perception* (Rastier 2010: Sect. VIII) can explain the identification and the management of the *semantic forms*. This notion represents also a synthesis of Rastier's approach to semantic analysis: to the cognitive perspective supposing the meaning subordinate to the concept and its compositionality, he opposes a structural and hermeneutic one, first kept close to the *Gestaltpsychologie* (Rastier 2010: 206). This approach reveals to be both *gestaltist* and *hermeneutic*, implying that the local dimension is determined by the global one, and supposing that the semantic analysis seems more like a process of identification of forms than like a process of computation (Rastier 2010: 207).

2.2.3 Instruction: *A Dynamical and Topological Notion*

The notion of *instruction* in linguistic analysis comes from different epistemological paradigms. It is used by many linguists, from Harris (1954a, b) to Fauconnier (1997) and more recently Tyler and Evans (2003) or Harder (2009). French linguists like Ducrot (1984), Anscombre and Ducrot (1983), or Nemo (2001) use it as well. We encounter it in the field of cognitive psychology (Barsalou 1999, 2008, 2010) as well as in the pragmatic tradition (Nølke 1994; Moeschler 2005).

According to Weinrich (1976) who proposes a "*C-I-T [Communication—Instruction—Text] linguistics*", texts are "communicative units" and *communication* is essentially a praxis whose pragmatic value is condensed in the notion of *instruction* (Weinrich 1976, Chap. IV) considered as an order or a suggestion produced by a subject to induce his interlocutor to react anyway, even only understanding the text. The text considered as *texture*—made by a plurality of connections in which its components are involved—suggests to the interpreter the different steps of his interpretative action, offering to him some instructions to follow and go through. Furthermore, it has a continuous communicative relation with a *context*, so its analysis begins identifying big units composing the "text-in-context" delimited from some evident interruptions of communication allowing to consider the text itself as an autonomous unit (Weinrich 1976: 15–16).

Considered as a *texture,* the text is an *actual system* of connections, which realizes partially the possibilities offered by language considered as the *virtual system*. This process of actualisation includes the selection of different options and their presentation as a structure: «A structure treated by processing as a single unit can be termed a *chunk*, and may range from a *local* (small-scale) *micro-structure*, over to a *global* (large-scale) *macro-structure*. A chunked micro-structure constitutes a *micro-state* of the actual system, while a chunked macro-structure constitutes a *macro-state*» (De Beaugrande 1980, *on line*). Linguistic elements becomes *instructions* for the actualisation: speakers need to know some *procedures* to have access to the resources offered by the linguistic system, as well as in its syntactic and semantic dimensions, supposing two kinds of knowledge: assertive and procedural. So each linguistic element becomes an instruction for the interpretative work made by the speaker, «e.g. a word being a pattern of sounds, a piece of a phrase, an instruction to 'activate' a meaning, and so on» (De Beaugrande 1996: 11), as an intermediate step for a more general *instruction of sense* (Cf. Segre in Weinrich 1976).

In this model, Weinrich adopts a dynamical-pragmatic concept of linguistic sign: it is an instruction to act, supposing an intransitive concept of meaning. So, the *semantic value* of a linguistic sign consists in showing to the interpreter what he has to do for interpreting and understanding. The model of communication is dynamical: every linguistic sign is understood as an instruction for the interpreter to act in a particular context[3] (Weinrich 1976).

So in the text conceived as a whole, linguistic signs become strategic components of its texture. In fact, they produce an expectation of sense which can be confirmed or denied by the text itself and by which the listener or reader develops his interpretative work (Weinrich 1976). Studying grammar in relation with the process of learning of foreign languages, Oller (1979) describes a phenomenon called *Expectancy Grammar*, consisting in the ability of the interpreter of a foreign language to make plausible hypothesis of what will be said or read in a specific context. The Expectancy Grammar depends on three parameters: (1) the awareness of situations and contexts, which makes anticipate the uses and the goals attended in communication; (2) the redundancy, that is to say supplementary information given by contexts, co-texts and paratexts; (3) the encyclopaedic knowledge, by which the interpreter creates some hypothesis on what will be said or written and solve the ambiguity of sense.

Semantic and syntactic information delivered by linguistic signs in relation with other linguistic signs belonging to a same linguistic system produce some expectations of sense in relation with the other textual components. Then Weinrich identifies three kinds of *preliminary informations* given, respectively, by (1) language; (2) text; (3) context. These ones represent the first step to advance in global understanding of text. For instance, according to Weinrich, we can consider the title as a *macrolinguistic instruction to the expectation* (Weinrich 1976; our emphasis). So, text becomes a global structure generated by the connection of different local structures whose connections make the *textuality* of the text (Weinrich 1976). Finally, all linguistic signs are for the reader *local instructions* whose connections give a global instruction of sense of the text considered as a whole (Weinrich 1976).

Referring to *textual pragmatics*, we have to suppose a concept of text defined as a *communicative unit*. Considering the relation between the linguistic and the extra linguistic contexts, the speaker and the reader, the supposed encyclopaedic

[3]Weinrich adopts the classical communicative model proposed by Shannon and Weaver (1949).

knowledge and the will of the interpreter to cooperate in communication (Bar-Hillel 1970: 270–271), also the influence of textual pragmatics in Eco's theory of text is condensed in the notion of instruction. For instance, he proposes to conceive terms as «instructions oriented to the text» (Eco 1979: 15), referring to an *Instruktionssemantik* textually oriented.[4]

The specificity of the notion of *instruction* in compositionnal gestaltist semantics is that it implements the two step process of convocation-evocation and that it accounts for the assembly of the units of an utterance. Furthermore, an instruction is unique and has various effects. Eventually, an instruction is underspecified but the meaning of the unit enriches in the course of discourse.

2.3 Continuity, Instruction and the Online Processing of Meaning

2.3.1 Two Fundamental Hypotheses Concerning Meaning

Following the theoretical background exposed in the previous section, our analysis is based on two main hypotheses:

- Individual words do not correspond to a list of word senses. Instead, we think that meaning emerges from the interaction with the context. We thus adopt an 'emergentist' approach that radically contrast with traditional approaches defining the different meanings of a word a priori, generally through a structured list of entries stored in the dictionary.
- Another related hypothesis is that there are not clear-cut sense boundaries. Instead, we consider meaning as a continuous phenomenon that emerges dynamically from all the interactions of the different linguistic units within a sentence (and, more generally, within the context as it may also include non linguistic elements).

Both hypotheses need a thorough analysis, involving different kind of utterances so as to explore their consequence. For example, the first hypothesis considers meaning as an emergentist phenomenon but does this entail that linguistic units do not have any meaning per se? If this is the case, how could one give the definition of a word out of context? Or alternatively, if each word has a core meaning, how can we define this core meaning? As for the second hypothesis, what is the nature of continuity in linguistics? Is it just a metaphor? Does it involve a

[4]Cf. in the text by Eco (1979), from Schmidt (1976: 56): «un lessema si può concepire teoricamente come una regola (in senso ampio) o un'istruzione per la produzione di un dato "comportamento" verbale e/o non verbale… Il campo-contesto [il campo lessematico] assegna al lessema le sue possibilità generali di funzionamento nei testi».

multidimensional analysis? In other words, is meaning directly related to continuity or does this continuity correspond to the composition of binary functions for example?

In the next section, we address some of these issues by considering practical examples in English, especially phrasal verbs and the case of "over", since these items are known to be hard to categorize in English.

2.3.2 The Case of English Complementation

Phrasal verbs are especially challenging in English. There has been numerous discussions to try to determine whether phrasal verbs can be decomposed or not (i.e. if the meaning of the phrasal verbs can be derived from the meaning of the base verb combined with the meaning of the particle). Of course, it all depends on the phrasal verbs under consideration: some phrasal verbs seem more compositional than others (if it makes sense to consider degrees of compositionality). In fact, the problem is not straightforward since the notion of phrasal verbs itself is a difficult and not so well defined one. Base verbs can be complemented with words that are sometimes considered as particles (when they are closely linked to the verb), sometimes as prepositions (when the word seems more closely related to the complement noun phrase). You can for instance "give up linguistics" but also "walk up the stairs": the word "up" plays a different part in each phrase, whether it is considered as a particle or as a preposition, depending on its relations with the base verb.

Things are probably clearer if we have a closer look at an example. Let us consider the following phrases:

- *Run into the room*: *into* introduces a location (where?), cf. "enter the room";
- *Run into financial difficulties*: *into* introduces the topic (what?), cf. "encounter difficulties"
- *Run into a telephone pole*: *into* introduces the topic (what? but not where?), cf. "collide with a pole"
- *Run into John*: *into* introduces a person (who? but not where?), cf. "encounter John".

From this example, one can see that the meaning largely depends on the noun that is involved: its nature (location, person or object) has a major influence on the meaning of the verb. The so called literal (or prototypical) meaning of *to run* (*to walk fast*) just reflects, in our opinion, a tendency to favour concrete meaning over more abstract ones. Most cognitive linguists also defend the fact that literal meaning comes first mostly based on evidence from complexity and learning theory. However, if one believes in usage and statistical significance, the supremacy of literal meaning does not hold and is even not justified any more. Instead, we observe that abstract meaning is prominent (in "*run into an obstacle*", *run* does not correspond to *walk fast* and the noun is not a physical object). Depending on the nature of the noun and the verb, into can be categorized as a preposition or a particle.

2.3.3 *The Specific Case of Over*

Let us take an example and let us now examine the case of *over*. *Over* is known to be plurifunctional and highly polysemous and hence has been extensively studied by linguists, especially in the cognitive linguistics tradition, see Taylor (1988), Deane (1993), Dewell (1994), for the last century, and Deane (2005), Talmy (2005), Tyler and Evans (2001, 2003), Van der Gucht et al. (2007) or Zlatev (2003) more recently. We have proposed our own study of *over* in (Col and Poibeau 2014) where we have shown that patterns of usage can be derived from corpora.

The following table provides an overview of the main patterns that can be derived from a corpus analysis:

N_{process}	$\text{over}_{\text{prep adjunct}}$	N_{temporal}
$N_{\text{cognitive}}$	$\text{over}_{\text{prep argument}}$	N_{topic}
N_{control}	$\text{over}_{\text{prep argument}}$	N_{topic}
V_{process}	$\text{over}_{\text{prep adjunct}}$	N_{temporal}
	$\text{over}_{\text{adverb}}$	$Q_{\text{quantifier}}$
V_{process}	over part	(N)
V_{process}	$\text{over}_{\text{adverb and over adverb}}$	
$N_{\text{processual}}$	$\text{over}_{\text{adjective}}$	

Some examples seem quite easy and simple to analyse:

- "they are often doing speeds **over** 50 mph": *over* is here an adverb expressing the notion of 'excess';
- "he could see **over** the tops of the trees": preposition with a spatial meaning;
- "the idea of a black sheep, has changed **over** the years": preposition with a temporal meaning;
- "controversy **over** bluefin tuna": preposition expressing the topic;
- "in November 1989 he took **over** as CDU leader": particle expressing control.

These examples seem straightforward because of the high frequency of some patterns or constructions in corpora:

- N1 over N2 is a structure that favours the meaning of topic for N2 and thus entails the categorization of *over* as a preposition, esp. with N1;
- over $N_{\text{quantifieur}}$: favours the excess meaning of *over*;

- over $N_{Temporal}$ (over the years): favours the categorization of *over* as a preposition with a temporal meaning introducing a modifier[5];
- only a few verbs are regularly associated with *over* as a particle *same parks* and are lexicalized as such in lexical databases like Wordnet.

One can also observe more complex cases. For example in "*then they have to go through a gruelling contest over a period of time*", it is not clear whether "*period of time*" refers to a topic (in which case the complement is an argument) or just to a period of time (in which case the complement is a modifier). There is here a tension between "*over a period of time*" that is most of the time a modifier and "*contest over*", which is generally followed by a noun referring to the topic of the contest.

Nuances of meaning can also been found in examples like "*Craig ran over the neighbor's cat*" versus "*She ran over the fields, past the barn, and to the white house*". The nature of the complement of *over* entails a largely different interpretation of the scene described by each of these sentences.

By these examples, we hope to have made it clear that the meaning of a word like *over* cannot be reduced to one dimension and can hardly be defined in itself. Instead, it is the interaction with the context that makes it possible to justify:

- The continuity of meaning.
- The different categories *over* refers to;
- The different function of the phrase introduced by *over* (esp. argument or modifier if we adopt a syntactic point of view).

It is probable that these very regular and highly frequent patterns are registered as such in memory. Bolinger has shown as soon as in the 1970s that language is made of co-occurring words with different kinds of constraints. Bolinger (1968) proposes to call prefab local co-occurring units to encompass words, collocations and idioms. For Bolinger (1976) these prefab should play a major role in semantics, beyond just simple and isolated words. Construction grammars (Goldberg 1995) can be seen as going a step further since they propose to generalize the analysis at different levels, taking into consideration the different kinds of interactions at any level in the analysis (words, idioms, phrases, sentences).

This tradition has also been acknowledged in the domain of corpus grammar. Collocations and prefab are especially important for Sinclair (1991) and there are attested traces of the interest of Bolinger for corpus linguistics (Fortis 2014).

The previous paragraphs make it clear that the core meaning of a linguistic unit like *over* is to be understood and described in the complexity of its uses. So as to account for the various uses associated with syntactic patterns and categories, and the extremely large (or maybe unlimited) number of possible interactions of the unit with its co-text, Col et al. (2012) suggest a model based on a gestalt compositional approach. We consider that the meaning of an utterance is constructed in the course of a double, dynamic process leading to a temporary, tentative semantic balance

[5]In most studies, *over* has a temporal meaning in this kind of examples but it should be discussed if this is necessary, since the temporal meaning is mostly supported by the noun following *over*.

between the different interactions. This balance is also largely self-organized as it is the result of two opposed movements, a bottom–up one and a top–down one. Each unit interacts with its co-text, thus the meaning of each unit depends on the meaning of the other units with which it interacts, and the global meaning of the utterance retro-acts on the meanings of the units present in the utterance and the co-text. This approach is then both compositional and gestalt as it focuses on the dynamicity of the interactions in the construction of meaning.

From this viewpoint, the meaning of a unit does not have to encapsulate all its attested meanings. It does not have to express a very general, abstract meaning either. It is rather its dynamicity which should be its core meaning and more generally of any linguistic units, and this dynamicity should be expressed according to what Victorri and Fuchs (1996) calls a convocation-evocation principle. The double movement we described above actually corresponds to two steps of the construction of meaning. The first step determines what needs to be present in the co-texte and more generally in the intersubjective field so that the unit can play its role in the construction. These elements that are necessary but not supplied by the linguistic unit itself are said to be "convoked". The second step refers to what the unit brings to the construction when interacting with other linguistic units. This interaction has an effect on the verbal scene under construction: this is what the unit "evokes". This capacity of convocation-evocation is what defines a linguistic unit; it is not its "meaning" per se but what we call its "instruction". As an example, the instruction supplied by *over* will be the following one (Col 2012; Col and Poibeau 2014):

> OVER convokes a bounded domain and evokes a movement of covering of the domain, including its bounds.

As can be noticed in the formulation of this instruction, no particular meaning is privileged and no proto-meaning either. The bounded domain is neither spatial nor temporal. It does not correspond to the use of *over* as a preposition or as an adverb but it is compatible with these various categories. In an example such as: "*Go over the bridge and turn right immediately onto a track leading into the trees*", it is clear that the "bounded domain" is a spatial one: it construes the space under the bridge (a river or a road) that is minimally delimited by both ends of the bridge itself. In "*As he released his anger, he felt his love for his ex-wife, and wept over their divorce.*" The domain is not spatial but still *over* convokes a bounded domain, the divorce, whose boundedness comes from the specificity of the divorce, i.e. their divorce. It also comes from the imperfectivity construed both by the tense of the verb "weep" (*wept*) and by the morpheme <ex-> in *ex-wife*. This last example is particularly interesting as we clearly notice that it is in the interaction with *weep* and *divorce* that *over* succeeds in giving the meaning of "topic" to the sentence. Still, *over* alone does not make the Topic meaning emerge; it needs the other lexical units to construe it, and particularly an 'object to weep over'. The interaction between *over* and the lexical units of the utterance (as well as its recurrence in regular patterns like N_1 $\text{over}_{\text{prep}}$ N_2) facilitates the emergence of the Topic meaning.

Finally, the dynamic aspect of the formulation aims at anticipating the assembly of the unit with the neighbouring ones. We enlarge on this last point now.

2.3.4 *An Instruction at Work: The Processing of the Instructions of an Utterance*

The processing of a complete utterance follows the same instructional principles: all the units are processed according to their instructions in the unfolding of discourse and when they are perceived. More precisely, Col et al. (2012: 161) suggest four main principles controlling the processing of the utterance, and above all, controlling the order of processing: (1) the units are taken into account one after the other as soon as the sentence is perceived; (2) once they are taken into account, and if it is possible for them to be processed, they are processed at once in conformity with their instruction. Thus, each unit plays its evocation role as soon as the elements it convokes are available; (3) if they cannot be processed, they are left pending until the elements they convoke are introduced on the scene that is being built; (4) once evoked, the elements of the scene continue to be determined or even transformed by later convocations by other units.

Let us take a simple example so as to illustrate the whole process:

Britain consumed over 9 million tons of paper.

Once perceived, "Britain" can be processed rather rapidly. It evokes an entity identified as a proper name referring to a country. Nothing about its status as an argument or another syntactic function is supplied yet; it is just an entity directly perceived on the verbal scene. The same could be more or less said about "consumed". Concerning this unit, we should make a distinction between the lexeme and the morpheme <–ed>. "Consume" immediately evokes a process—there is no noun corresponding to this lexeme in the English lexicon. But the <–ed> morpheme may evoke at least two different elements: either a process located in the past time (or located on a different level from the actual one), hence the interpretation of <–ed> as a preterit, or an accomplished process, hence the alternative interpretation of <–ed> as the past participle. But, since an entity has been introduced on the scene (Britain) and as this entity has no real function yet, the <–ed> morpheme is processed as a marker of the preterit and at the same time, Britain acquires a syntactic function: subject of "consumed". Both elements are determined simultaneously in the course of their interaction and of their perception as a "good form", in a gestalt way.

The rest of the processing is more complex. Actually, when "over" is perceived, it cannot be processed because at this stage of the processing, it is not recognizable as a preposition or an adverb (nothing is associated with it except a verb and a subject), nor as a verb particle as "consume over" is not a phrasal verb. "Over" is then left pending until other units are processed. The following unit, "9" cannot be processed either: "Britain consumed over 9" is barely understandable and needs the expression of a quantity to gain some meaning. "Million" is the next unit to be processed as "9" is left pending and together, they evoke the quantity required by the process "Britain consumed". We may say that "9" attracts this quantity as soon as it is perceived. As for the object of the quantity ("tons of paper"),[6] it is rapidly processed insofar as a quantity plus a process are present on the scene and they can assemble.

At the stage of the processing, "over" is finally processed. This stage is a double one in fact: "over" is processed, i.e. a semantic trajectory is given, as it were, and at the same time, this unit is categorized, namely as an adverb thanks to the instruction supplied both by the verb and the nominal complex evoking a quantified entity. "Over" is the last unit to be processed and it evokes something which goes beyond the movement of covering supplied by its instruction. In the interaction with the rest of the utterance, its contribution is larger than identifying the meaning of a unit: it finalizes, at least tentatively, the semantic gestalt construed by the whole utterance.

2.4 Continuity in Language Modelling

Since the availability of numeric methods and very large corpora, a lot of progress has been made in the automatic exploration of lexical semantics. Most approaches rely on distributional methods, based on a hypothesis first formulated by Harris in 1954. This hypothesis states that words appearing in the same context tend to have a similar meaning (the same tradition was popular in linguistics, cf. Firth (1957): «You shall know a word by the company it keeps». The number of papers published on distributional methods is nowadays so large that it is hard to give a comprehensive picture of the field. However, most approaches so far were considering isolated words to calculate word similarity (for information retrieval or information extraction applications for example). A large effort has been made recently to go beyond the word barrier and take into account larger phrases and more generally the syntactic context so that distributional methods can cope with the word limitation problem.

The hypothesis that makes it possible to take into consideration sequences more complex that isolated word is the notion of compositionality, often attributed to Frege (1892). Compositionality means that the semantics of a sequence can be calculated from the semantics of its parts. In other words, it would be possible to combine the semantics of the parts to compute the semantics of a complex sequence. This is supposed to explain why humans can understand sentences they have never seen before: by combining the meaning of isolated words, we are able to

---

[6]We do not describe the full processing of this nominal complex but various elements at different levels like the morpheme (-s) or prepositions (of) are to be distinguished.

compute the meaning of larger sequences even if we have never seen these sequences before.

These methods are powerful and seem well fitted for language modelling. We will not detail here the mathematical basis for these models, which is can be quite complex. We think it is more relevant to focus on different features that make these models interesting from a linguistic point of view.

1. Thanks to distributional methods, it is not any more necessary to define in advance what is the meaning of a word. Distributional methods take literally the principle that words are defined by the contexts in which they appear, which lead to representations that are quite different from those of existing human dictionaries. Dictionary induction is a task examining to what extent the meaning of a word can be induced only by looking the context: results obtained with this approach are often interesting, sometimes debatable and at least question existing lexical descriptions.
2. The description obtained by such methods is consistent with the notion of "continuity" (Fuchs and Victorri 1994). One does not need to a priori define clear-cut word senses but different descriptions, with a different granularity, can be obtained depending on how precise the context taken into account is. It is thus possible to define more or less fine grained word senses for a same word, and propose semantic maps showing that certain senses are closer than others (see Fig. 2.1).

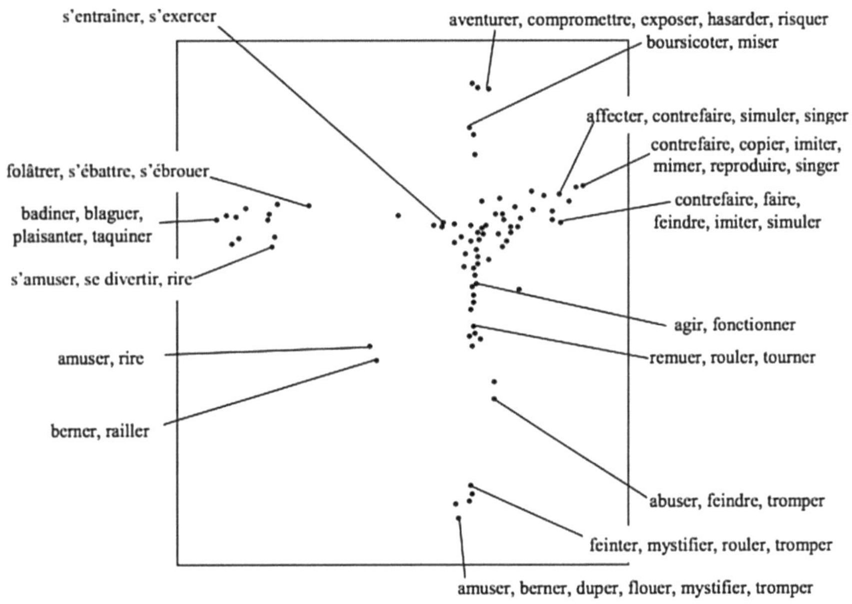

Fig. 2.1 The semantic space associated with the French verb jouer, from Gaume et al. (2005)

The techniques used are also able to compare different words so as to automatically identify synonyms or other related words, identify regular patterns of meaning changes (metonymy, metaphors, etc.), cf. Shutova et al. (2012).

3. When coupled with a way to deal with compositionality, distributional methods can be extended so as to calculate the meaning of phrases or even sentences (however, note that most models are limited due to the complexity entailed by pure distributional models. An open research question consists in extending these methods using tractable models). The advantage of applying these methods beyond the word limit is to represent using a uniform mechanism notions such as idioms, compounds or multi-word expressions. An idiom can be seen as a regular phrase with limited possibilities of variations, which can be to a large extent characterized and calculated automatically using statistical and/or distributional methods.

The complexity of nowadays techniques is high since distributional methods require to calculate the co-occurrences of words and phrases through very large corpora (sometimes gigabytes of data, several billions words). Since the current trend is to go beyond the word limit, there is potentially an infinite number of sequences to take into account, which quickly lead to combinatorial explosion and other complexity problems.

Different proposals have recently been made to implement this kind of models. A first approach consists in generalizing the use of matrices (that can represent the co-occurrences of a series of words) through tensors (Van de Cruys et al. 2013). The key idea is that compositionality is modelled as a multi-way interaction between latent factors, which are automatically constructed from corpus data. The composition component can itself be based on a formal description of interaction between linguistic items (Grefenstette et al. 2014). Another very popular way of dealing with the problem is to use multi-layered neural network, a method known as "deep learning", widely used initially in speech recognition and nowadays in natural language processing (Sarikaya et al. 2014; Socher et al. 2013). It seems that there is thus a real convergence between automatic methods and linguistic description here.

2.5 Conclusion

In this paper, we have considered the notions of continuity and interaction in linguistic models. There is now a quite rich tradition of work based on the hypothesis that natural language is not a discrete model. Instead, continuous models consider word sense, grammar rules and categories as continuous notions: some words are hard to categorize and some rules do not fully apply in certain contexts. Word meaning in context often corresponds to different meanings according to the dictionary, etc.

New models have been proposed that provide a more accurate view of these phenomena. Recent advances in natural language processing also support this idea by providing very rich models based on a multi-level representation of the context of use of linguistic items. These models have been applied to different problems (from lexical categories to grammar rules, but also to more complex tasks such as information extraction or machine translation). From this point of view, we can see an interesting convergence between linguistic and computational models.

There is anyway still a long way to go in this domain. The models considered here are just in their infancy, especially for linguistic descriptions. What would be a possible perspective is the generalization of the linguistic description of a language from an instructional point of view. The purpose would be the establishment of a repertoire of instructions whatever their category may be, and the setting up of assembly rules in the formulations of these instructions. This kind of repertoire could then be used in the field of computer sciences and natural language processing.

References

Anscombre, J. -C., & Ducrot, O. (1983). L'argumentation dans la langue, Mardaga, Coll. Philosophie et langage, 184 p.

Bar-Hillel, Y. (1970). Communication and argumentation in pragmatic languages [1968], in Aa. Vv., *Linguaggi nella società e nella tecnica*, Milano, Comunità.

Barsalou, W. L. (1999). Perceptual symbol systems. *Behavioral and Brain Sciences, 22*, 577–660.

Barsalou, W. L. (2008). Grounded cognition. *Annual Review of Psychology, 59*, 617–645.

Barsalou, W. L. (2010). Grounded cognition: Past, present, and future. *Topics in Cognitive Science, 2*, 716–724.

Benveniste, E. (1966). *Problèmes de linguistique générale*. Paris: Gallimard.

Bolinger, D. (1968). *Aspects of language*. New York: Harcourt, Brace, and World.

Bolinger, D. (1976). Meaning and memory. *Forum Linguisticum, 1*, 1–14.

Col, G. (2012). Over the top: Quelle instruction sémantique pour l'unité *over*?. *Revue de Sémantique et de Pragmatique, 31*, 41–66.

Col, G., & Poibeau, T. (2014). An instruction-based analysis of over. *Language and Cognition, 6*(3), 370–407.

Col, G., Aptekman, J., Girault, S., & Poibeau, T. (2012). Gestalt compositionality and instruction-based meaning construction. *Cognitive Processing, 13*(2), 151–170.

Col, G., Aptekman, J., Girault, S., & Victorri, B. (2010). Compositionnalité gestaltiste et construction du sens par instructions dynamiques. *CogniTextes, 5*. URL: http://cognitextes.revues.org/372.

De Angelis, R. (2014). Il testo conteso. Semiotiche ed ermeneutiche nella seconda metà del Novecento, Pisa, ETS.

Deane, P. (1993). *Multimodal spatial representations: On the semantic unity of 'over' and other polysemous Prepositions*. Duisburg: Linguistic Agency, University of Duisburg (LAUD).

Deane, P. (2005). Multimodal spatial representation: On the semantic unity of over. In: B. Hampe, & J. E. Grady (Eds.), *From perception to meaning: Image schemas in cognitive linguistics* (pp. 235–282). Berlin/New York: Mouton de Gruyter.

De Beaugrande R. A. (1980). Text, discourse and Process. Toward a Multidisciplinary Science of Texts. In *Advances in discourse processes* (vol. XI). Norwood (New Jersey): Ablex Publishing Corporation.

De Beaugrande R. A. (1996). New foundations for a science of text and discourse. Cognition, communication and the freedom of access to knowledge and society. In *Advances in discourse processes* (vol. LXI). Norwood (New Jersey): Ablex Publishing Corporation.
de Saussure, F. (2002). *Écrits de linguistique générale*. Paris: Gallimard.
de Saussure, F. ([1916] 1922). *Cours de linguistique générale*, C. Bally et A. Séchehaye éds., avec la collaboration de A. Riedlinger, Paris, Payot.
Dewell, R. B. (1994). Over again: Image-schematic transformations in semantic analysis. *Cognitive Linguistics*, *5*, 351–380.
Ducrot, O. (1972). *Dire et ne pas dire*. Paris, Herman: Principe de sémantique linguistique.
Ducrot, O. (1984). *Le Dire et le dit*. Paris: Editions de Minuit.
Eco, U. (1979). *Lector in fabula*. Milano, Bompiani.
Eco, U. (1988). Sémiotique et philosophie du langage, Paris, PUF [first edition: 1984, Semiotica e filosofia del linguaggio, Torino: Einaudi].
Erk, K., & McCarthy, D. (2009). *Graded word sense assignment. Proceedings of the Empirical Methods in Natural Language Processing*. Singapore.
Fauconnier, G. (1997). *Mappings in thought and language*. Cambridge: Cambridge University Press.
Firth, J. R. (1957). A synopsis of linguistic theory 1930–55. In *Selected papers of J. R. Firth (1952–59)* (pp. 168–205). London: Longmans.
Frege, G. (1892). *Über Sinn und Bedeutung*. ¨Zeitschrift fur Philosophie und philosophische Kritik, *100*, 25–50.
Fortis, J. -M. (2014). La linguistique cognitive comme réseau théorique. *Colloque SHELD-HTL*.
Fuchs C., & Victorri, B. (Eds.). (1994). *Continuity in linguistic semantics*. Amsterdam; Philadelphia: J. Benjamins Publishing.
Gaume B., Venant F., & Victorri B. (2005). Hierarchy in lexical organisation of natural languages. In D. Pumain (Ed.), *Hierarchy in natural and social sciences* (Vol. 3, Methodes series, pp. 121–143). Springer.
Girault, S., & Victorri, B. (2009). Linguistiques de corpus et mathématiques du continu. *Histoire, Epistémologie, Langage, 31*(1), 147–170.
Goldberg, A. (1995). *Constructions*. Chicago: The University of Chicago Press.
Grefenstette E., Sadrzadeh M., Clark S., Coecke B., & Pulman S. (2014). Concrete sentence spaces for compositional distributional models of meaning. In H. Bunt, J. Bos & S. Pulman (Eds.), *Computing meaning* (Vol. 4, pp. 71–86). Springer.
Guillaume, P. (1979). *La Psychologie de la forme*. Paris: Flammarion.
Harder, P. (2009). Meaning as input: The instructional perspective. In V. Evans & S. Pourcel (Eds.), *New Directions in Cognitive Linguistics*, 15–26.
Harris, Z. S. (1954a). Distributional structure. *Word, 10*(23), 146–162.
Harris, Z. S. (1954). Transfer grammar. *International Journal of American Linguistics, 20*(4), 259–270.
Köhler, W. (2000). *Psychologie de la forme. Introduction à des nouveaux concepts en psychologie*. Paris: Gallimard.
Manning, C. (2003). Probabilistic syntax. In R. Bod, J. Hay & S. Jannedy (Eds.), *Probabilistic Linguistics* (pp. 289–341). MIT Press.
Maturana H. R., & Varela F. J. (1980). *Autopoiesis and Cognition. The Realization of the Living*. Dordrecht, Holland/Boston, U.S.A./London, England: D. Reidel Publishing Company. (or. ed. De Maquinas y seres vivos, Editorial Universitaria S. A., Chile, 1972).
Moeschler, J. (2005). Connecteurs pragmatiques, inférences directionnelles et représentations mentales. *Cahiers Chronos*, *12*, 35–50.
Nemo, F. (2001). Pour une approche indexicale (et non procédurale) des instructions sémantiques. *Revue de Sémantique et de Pragmatique, 9–10*, 195–218.
Nølke, H. (1994). *Linguistique modulaire: de la forme au sens. Bibilothèque de l'Information Grammaticale*. Paris-Louvain: Peeters.
Oller, J. W. (1979). *Language tests at school: a pragmatic approach*. London-New York: Longman.

Petitot-Cocorda, J. (1985). *Morphogenèse du sens*. Paris: PUF.
Rastier, F. (1989). *Sens et textualité*. Paris: Hachette.
Rastier, F. (1997). Herméneutique matérielle et sémantique des textes. In J. -M. Salanskis, F. Rastier, & R. Scheps (Eds.),*Hermémenutique: Textes, sciences* (pp. 119–148). Paris: PUF.
Rastier, F. (2001). Du signe aux plans du langage. *Cahiers Ferdinand de Saussure, 54*, 177–200.
Rastier, F. (2006). Formes sémantiques et textualité. *Langages, 163*, 99–114.
Rastier, F. (2009). *Sémantique interprétative*. Paris: PUF.
Rastier, F. (2010). *Sémantique et recherches cognitives* (3a ed.). Paris: PUF.
Sarikaya, R., Hinton, G. E. & Deoras, A. (2014). Application of deep belief networks for natural language understanding. *IEEE Transactions on Audio, Speech and Language Processing*.
Schmidt, S. (1976). *Texttheorie*. München: Fink.
Shannon, C., & Weaver, W. (1949). *The mathematical theory of communication*. Urbana Illinois: The University of Illinois Press.
Shutova, E., Van de Cruys, T., & Korhonen, A. (2012). Unsupervised metaphor paraphrasing using a vector space model. In *Proceedings of COLING 2012, Mumbai, India*.
Sinclair, J. (1991). *Corpus, concordance, collocation*. Oxford University Press.
Socher, R., Manning, C., & Bengio, Y. (2013). Deep learning for NLP (without magic). In *Tutorial Association of Computational Linguistics 2013* (http://www.socher.org/index.php/DeepLearningTutorial/DeepLearningTutorial).
Talmy, L. (2005). The fundamental system of spatial schemas in language. In B. Hamp (Ed.), *From perception to meaning: Image schemas in cognitive linguistics* (pp. 199–234). Mouton de Gruyter.
Taylor, J. R. (1988). Contrasting prepositional categories: English and italian. In: B. Rudzka-Ostyn (Ed.),*Topics in cognitive linguistics*. John Benjamin.
Thom, R. (1972). *Stabilité structurelle et morphogenèse*. Paris: Interéditions.
Thom, R. (1974). *Modèles mathématiques de la morphogenèse, Collection 1018*. Paris: Union Générale d'Éditions.
Tyler, A., & Evans, V. (2001). Reconsidering prepositional polysemy networks: The case of over. *Language, 77*, 724–765.
Tyler, A., & Evans, V. (2003). *The semantics of english prepositions: Spatial scenes, embodied meaning and cognition*. Cambridge: Cambridge University Press.
Van de Cruys, T., Poibeau, T. & Korhonen, A. (2013). A tensor-based factorization model of semantic compositionality. In *Proceedings Human Language Technologies: Conference of the North American Chapter of the Association of Computational Linguistics, Atlanta, Georgia* (pp. 1142–1151).
Van Der Gucht, F., Willems, K., & De Cuypere, L. (2007). The iconicity of embodied meaning. Polysemy of spatial prepositions in the cognitive framework. *Language Sciences, 29*, 733–754
Victorri, B., & Fuchs, C. (1996). *La Polysémie*. Paris: Hermès.
Weinrich, H. (1976). *Sprache in texten*. Stuttgart: Enrst Klett.
Zlatev, J. (2003). Polysemy or generality? Mu. In: H. Cuyckens, R. Dirven, & J. Taylor (Eds.), *Cognitive approaches to lexical semantics* (pp. 447–494). Berlin/New York: Mouton de Gruyter.

Chapter 3
Modeling Language Change: The Pitfall of Grammaticalization

Quentin Feltgen, Benjamin Fagard and Jean-Pierre Nadal

3.1 Introduction

Since the 1990s, understanding the mechanisms underlying language emergence and evolution is the subject of many studies (Christiansen and Kirby 2003; Fitch 2005; Steels 2011). Yet, it is not an easy task to account for these changes: timescales of interest are overlapping (Croft 2013), data are often sparse and/or uneasily processed (how to meaningfully present data remains an open question), and above all, every insight one may have about these changes drastically depends on the ideas one has about language: these ideas constitute an unavoidable prism whose specific biases are imparted to the interpretation and the understanding of linguistic phenomena.

Several attempts to model these changes have been proposed; they are generally due to the merging of efforts from linguists and other scientists coming from fields where modeling has already become a habit: biologists, physicists, population

Q. Feltgen · J.-P. Nadal
Laboratoire de Physique Statistique, CNRS - Ecole Normale Supérieure - UPMC - Univ., Paris Diderot, PSL Research University, 24, rue Lhomond, Paris 75231 cedex 05, France
e-mail: quentin.feltgen@ens.fr

J.-P. Nadal
e-mail: nadal@lps.ens.fr

B. Fagard (✉)
Laboratoire Lattice, CNRS, Ecole Normale Supérieure, Université Sorbonne Nouvelle, PSL Research University, Sorbonne Paris-Cité, 1 rue Maurice Arnoux, Montrouge 92120, France
e-mail: benjamin.fagard@ens.fr

J.-P. Nadal
Centre d'Analyse et de Mathématique Sociales, CNRS-Ecole des Hautes Etudes en Sciences Sociales, PSL Research University, 190-198 av. de France, Paris 75244 cedex 13, France

F. La Mantia et al. (eds.), *Language in Complexity*,
Lecture Notes in Morphogenesis, DOI 10.1007/978-3-319-29483-4_3

geneticists, economists, mathematicians. Such models combine linguistic hypotheses on language and language change with a quantitative formulation, which may lead to numerical simulations or mathematical considerations (Vogt and De Boer 2010). They may even lead to implement algorithms of language learning and communication in artificial intelligence, as has been done for the Sony robots (Steels 1999; Steels and Belpaeme 2005).

Two main directions have been thoroughly explored so far: one focusing on the social, conventional nature of language, mainly addressed with agent-based models, and one relying on transformational grammar, where language is seen as a set of rules learned by children, whose mistakes result in language change. However, neither manages to successfully model grammaticalization, although this phenomenon may be regarded as one of the major mechanisms of language change and language evolution (Meillet 1912; Peyraube 2002).

Grammaticalization (Meillet 1912; Hopper and Traugott 2003) is the process whereby a word (or any other linguistic element) acquires new grammatical functions and grammatical characteristics. It has been thoroughly explored and detailed in numerous works (cf. Prévost 2003 for a discussion, and Heine 2003 for an overview).

Nonetheless, both modeling approaches have proven rich and enlightening: our goal is merely to stress their limitations, and ideally to shed some light on the dynamics of grammaticalization phenomena with a more adequate model. We will first briefly review some of the works which have been proposed to model language change; these approaches will be then discussed and we will try to understand why they fail to satisfactorily account for grammaticalization. From what we will have learned, we will derive a few principles which may serve as a guideline in the modeling task. At last, we will lay the basis for a proposal of a model which, we hope, will account for some essential features of grammaticalization.

3.2 Two Roads to Language Change

Despite the fact that the field is very young, the literature concerning language change models is already impressive, both by its size and its diversity. It is not our purpose here to review in detail this polymorphic body of literature. In this section, we only aim to briefly recall the main paths which have been explored in these attempts to model language change. More extensive reviews can be found elsewhere (Castellano et al. 2009, Sect. V; Loreto et al. 2011).

We choose to present these different works as a diptych, where each plate is characterized by the representation of language on which these models rest: in the first series of approaches, language is seen as a set of sign-meaning pairs, i.e., a code (henceforth 'code' models); in the second series, language is seen as a set of rules (henceforth 'rule-based' models). Each of these two representations may vary a lot from one paper to the other; yet, we think that they both unify a sufficient number of works, and that they also differ enough from one to another to provide an efficient discrimination between the two, thus leading to a clear, unambiguous setting of the hinge which separates them.[1]

3.2.1 Code Models

In this first line of models we will review, language is seen as a code, that is, an association between forms (or signs, words, signals, sounds, etc.) and meanings (or linguemes, objects, events, etc.). The underlying assumption of this view is that meanings are clearly identified: they correspond to obvious, prominent features of the environment of the speakers; they are also shared by all speakers.[2]

3.2.1.1 Population-Based Code Models

In these models, the language of agents is characterized, for each user, by two matrices, often labeled P and Q: one is the coding matrix; the other is the decoding one. The production (or active) matrix P relates meanings to forms; the comprehension (or passive) matrix Q relates forms to meanings. Thus, when a speaker wants to communicate a given meaning to a hearer, he utters a form according to the specifications of the production matrix; the hearer, receiving this uttered form, uses his comprehension matrix to decode the form and recover the conveyed meaning. The size of these matrices (i.e., the number of meanings and forms) is the same for all agents.

This coding/decoding process can be either probabilistic or determinist. If it is determinist (Pawlowitsch 2007; Pawlowitsch et al. 2011), then only one form is associated to each meaning in the production matrix (excluding thus synonymy); simultaneously, only one meaning is associated to each form in the reception matrix (excluding polysemy and homonymy).

These models may also account for a protogrammar (Nowak and Krakauer 1999), where meanings are separated into distinct classes, objects, and actions: utterances then amount to the production of an object-action pair.

This framework often leads to the use of population dynamics. A population is described by the number of its individuals speaking each language (for instance, if there are three languages or linguistic variants L_1, L_2, L_3 in the population, this population will be described as the vector (x_1, x_2, x_3) of the proportions of individuals speaking each of these three languages). If an agent uses a language which allows him to communicate successfully with his fellow agents, then he will be able

[1]Note that some models are hard to classify; for instance, the Iterative Learning Model (Kirby and Hurford 2002), while closer to transformational grammar in its approach, falls into the first category.

[2]See however (Victorri and Fuchs 1996) on how such links may evolve and generate polysemy.

to reproduce more effectively, leading to the production of an offspring speaking the same language. Language thus plays a role similar to that of genome in population genetics. It should be stressed, however, that population dynamics do not describe each individual communicative event between every pair of agents; rather, they consider the averaged communicative potential of different languages.

3.2.1.2 Agent-Based Models

Similar dynamics may also be interpreted on the timescale of a human being (Cucker et al. 2004): agents change their language in order to be able to communicate more successfully with the rest of the population. Models adopting this interpretation may not strongly differ in their mathematical formulation from the previous models; yet, they lead to a very different vision of language evolution, since language changes now within individuals, rather than changing as individuals are replaced.

Most models based on the agent approach are utterance models. Through linguistic interactions, agents record all occurrences of forms expressing a given lingueme in the interactions in which they took a part. They, thus keep in memory in which proportion each different form has been uttered to express the lingueme, rather than the simple fact that such forms have been used in the past to express it. Such an approach is prototypical of the “exemplar” models introduced in phonology (Pierrehumbert 2001). Worth mentioning is the utterance selection model (Baxter et al. 2006) for which a detailed mathematical analysis has been done.

One of the most thoroughly studied models in language change is the naming game. It consists of an agent-based model where all agents cooperate to find an overall convention about meaning-forms association. In this type of model, the inventory of each agent is specified, and each of their interactions is accounted for. If the mechanisms governing the interaction between agents are simple enough, these agent-based models may be turned into population dynamics model (Castelló et al. 2009).

The naming game was first developed by Luc Steels to develop an efficient way of implementing a communicative behavior in robots (Steels 1995). Since then, it has been more extensively studied by the team of Vittorio Loreto (Baronchelli et al. 2006, 2008). Each agent is characterized by the set of inventories (i.e., list of possible names) it associates with each object. These inventories are modified as agents interact with each other, until all agents agree on the name to be given to each object. An interesting point is that there is no real difference between the naming of a single object and the naming of a plurality of objects, the latter option only leading to a slowing down of the conventionalization process.

Several extensions have been made around this naming game. It is possible to create new specific forms to make the mutual understanding easier (Omodei and Fagard 2013). The category game (Puglisi et al. 2008) considers a continuum of objects (the different light frequencies of visual stimuli) which is both perceptually and linguistically categorized.

To conclude this first subsection, it should be stressed that a majority of these models are devised to study how a linguistically heterogeneous population may reach a consensus, i.e., how all individuals may happen to speak the same language, through their successive interactions for agent-based models, or because one of the linguistic variants yields a higher fitness in population-based models. They generally address the question of the emergence of language, rather than the question of language change.

3.2.2 Rule-Based Models

While, the previous models focus on the lexical component of language, rule-based models approaches insist on the grammatical and structural rules. Language is described as a set of parameters (Chomsky 1981) specifying these rules (I-language), which in turn govern the production of utterances (E-language).

In this framework, language is seen as irreducible to biological evolution. The notion of language fitness does not play any major role in language evolution; however, as in most population models, language change is not usage-based, but results from the passing of human generations. More precisely, the source of the change is to be found in the process of language acquisition (Niyogi 2009): «Perfect language acquisition would imply perfect transmission. Children would acquire perfectly the language of their parents, language would be mirrored perfectly in successive generation, and languages would not change in time. [...] Therefore, for languages to change with time, children must do something differently from their parents».

This strong hypothesis on the origin of language change shifts the focus to this process of language acquisition. In all models exploring this process, a population is considered, and this population is split into two groups: adults and children. Adults have a fixed I-language, and produce utterances; children listen to these utterances, and infer the most plausible I-language compatible with the set of utterances they have been given.

Children have a limited time to learn language; once this time is elapsed, they definitely adopt an I-language (Kirby and Hurford 2002). The very existence of this time limit implies that children are exposed to a finite number of utterances—which is crucial: it is this finiteness which makes the acquisition process unavoidably imperfect.

On the basis of the utterances they have recorded during their learning period, children infer hypotheses to decide which I-language is the most compatible with these utterances. Such a statement raises an immediate question: from which set must an I-language be chosen? Rule-based models approaches make the following assumption: a set of possible I-languages innately exists in the mind of children; an I-language is chosen among this set of possibilities; such a set is called the Universal Grammar. Interestingly, this Universal Grammar is sometimes considered to be subject to biological evolution (Nowak et al. 2001).

A child is expected to learn an I-language close enough to the I-language of the adults. Otherwise, an individual would be able to communicate with neither individuals of other generations nor individuals of its own generation, which is obviously not the case. The question is then: what constraints should there be on the set of learnable I-languages, for a child to be able to learn an acceptable language? The answer to this question depends on the particular inferential procedure actually used by children; yet, no matter the procedure, this set of I-languages has to be finite (Nowak et al. 2001).

The rule-based models framework offers many more ways to explore language acquisition. Five possible lines of investigation have thus been proposed: «(1) the type of learning algorithm involved; (2) the distribution of the input data; (3) the presence or absence of noise or extraneous examples; (4) the use of memory; (5) the parameterization of the language space itself (the class of possible grammars/languages)» (Niyogi and Berwick 1996). If lines (1), (4) and (5) concern the cognitive process of inference, line (2) and more especially line (3) partly result from the specific social structure of the population. Those questions are indeed of some importance: do the children only listen to their parents, and so hear utterances produced on the basis of a very limited set of different I-languages? Does child—child communication have any influence on the learning process?

3.3 The Pitfall of Grammaticalization

These different types of models (code models and rule-based models) share some serious limitations if we want to address the question of why and how languages change, and especially, the question of grammaticalization. Indeed, grammaticalization intertwines the structure of the language and the particular role of its lexical elements: code models fail to account for the structure; rule-based models fail to describe the specific history of words and constructions. In the following, we will highlight three main limitations of existing models.

3.3.1 Communication

The first of these limitations concerns the communicative mechanisms at play in these models. Indeed, they are often very basic and we cannot expect them to give birth to interesting pragmatic constraints which would act on language change.

These communicative mechanisms are of two types. The first type is encountered in models such as the naming game. Agents try to find an agreement on the name associated with an object (sign-meaning pairing); they then check that there is no misunderstanding through nonverbal means. This latter mechanism has drastic consequences. It means that the language emerging from these games cannot be adapted to situations where one cannot point to the object referred to in its speech.

Two objections immediately arise: first, language is obviously much more complex, and can be used in a much wider range of situations; second, one can wonder why language would have been selected through the evolution process of mankind, if it had been no more efficient than a simple pointing device, which is found in other apes and even in lemurs (Gómez 2007; Leavens et al. 1996). Models such as the Naming Game are thus unable to describe the very essence of language. They have originally been created in a very specific goal (robotic implementation), for which such a nonverbal process has a meaning and a reality; but they might well fail to be extended to deeper linguistic investigations.

The second type escapes this limitation, since it does not refer to any nonverbal way to ensure communication. Let us consider a given context, sufficiently vague in its definition for an agent to be repeatedly exposed to it in a reasonably short time. At each exposure, the agent of interest is surrounded by agents who produce utterances. It may happen that, when exposed anew to a similar context, the agent of interest will himself produce an occurrence which will be more or less alike to what he has heard during the previous exposures. He may also consider what he has himself said in the past. Such a mechanism is at work in rule-based models (the child mimics what he has heard from adults), but also in a few agent models (as the utterance selection model, based on this very mechanism).

This last communicative mechanism seems to be more realistic. However, it does not explain why people produce utterances (the question of the function of language is thus still an open one), nor does it distinguish a situation in which people understand each other from one in which they just say similar things in similar situations, without even knowing why. And, above all, they do not take into account pragmatic considerations.

Yet, pragmatic constraints do seem to be an essential feature of language change (Croft 2013). Agent-based and population models show that a consensus is possible, and likely to occur; in several models, general consensus is even a stable attractor of the dynamics (Baronchelli et al. 2006; Kirby 2001). The question is then the following: if a society of human beings is able to reach a consensus concerning the conventions of language, why *do* languages change? Several hypotheses may be invoked: according to the invited inference theory, the repeated use of specific constructions in contexts implying further meaning may lead language users to associate this new meaning to the original meaning of the construction (Traugott and Dasher 2001); a desire to valorize one's own speech may exhort to use new, unusual constructions (Keller 2005); the subtleties of gallantry impose to constantly renew the courtesy formula, leading to a constant depreciation of past constructions (Keller 1989). In all these hypotheses, what drives change is pragmatics.

Even though simplifications and abstractions are the intrinsic preliminaries of any possible model, some of them can be misleading. The specificity of the linguistic phenomenon has been thoroughly argued for in the past; it is therefore crucial that models be able to account for this specificity. Putting apart pragmatic considerations in the communication rules actually implemented in a model may thus be slightly too bold a simplification.

3.3.2 Language

Another important problem of the existing models is the way language is implemented. We have already mentioned the two major representations encountered in the field: the code, and the set of rules.

In code models, language is viewed as but a set of sign-meaning associations; that is, it is no more complex in its structure than, say, the equivalence between Morse alphabet and Latin alphabet. This vision of language seems to be quite far from reality. It may be a useful simplification to study how a group of agents can reach a consensus, but it says only little on changes of linguistic nature.

Rule-based models offer an interesting alternative. In such models, the linguistic system is more elaborate: it is a set of parameters, or rules, which permit communication. For instance, one of these parameters can be a binary variable which is set to 0 if the language is of SVO type and to 1 if it isn't. The bit-string models (Schulze and Stauffer 2005; Zanette 2008), for instance, consider language as a bit-string, that is, a string of a fixed number of bits, each of them taking two values, 0 and 1. All of these bits indicate a dichotomic property of the language.

However, language, as it is actually used by speakers, is not simply an actuation of these parameters; indeed, such a view does not take individual words into account, words whose specific history seems to be an essential part of language change. One could also argue that all linguistic changes occur through the fortunes of individual words, and so it may seem that, by setting words aside, rule-based models also leaves aside (at least one of) the channels through which languages change.

Furthermore, rule-based models do not offer to test the hypothesis that transformational grammar is a relevant framework to study language evolution, since they are based on this very hypothesis. In other words, they take for granted the validity of their own framework. A more open position—as claimed in (Reali and Griffiths 2010)—would have been to construct a representation of language which may change both through children acquisition, and through other factors, to compare their importance, and to judge of their respective relevance. Thus, the predominance of language acquisition in the process of change could be more convincingly justified.

3.3.3 Change

The third line of discussion is possibly the most essential one, and partly follows from the two previous ones. In all these models, either language evolves toward a stable form, and does not subsequently deviate from it, or it evolves through extrinsic factors.

Agent-based models often fall into the first category[3]: they do not describe a language which is continually changing, but the emergence of a cultural agreement whose shape and modalities grasp some actual features of the language. Once such an agreement is met, then language keeps its final form and does not evolve further. This seems to be a consequence of the fact that language, in these models, is reduced to a pairing between two sets of abstract objects, and does not take into account any pragmatic tensions in the production of utterances.

In the majority of cases, whenever language does evolve, it does not evolve by itself, although it has been claimed elsewhere to be a self-organizing system (Bybee 2003; John and Bennett 1982): the evolution does not result from the regular, everyday use of the language, but is brought about by extrinsic factors (Wedel 2006). In rule-based models, language change results from changes in the population of speakers. The timescale of language change is thus given by the timescale of the passing of successive human generations—a timescale which does not necessarily coincide, however, with actual timescales of language evolution. In other models, change can come from the mixing of two different populations speaking different languages, or from speakers who cannot, or refuse to, follow the established convention (Komarova and Jameson 2008).

These different types of models thus all fail to give an account of language evolution in which language use actually triggers the changes at hand: language evolution is but an epiphenomenon, resulting from other phenomena occurring in the society of the speakers, and it cannot occur without these external perturbations. We prefer to assume that languages are intrinsically dynamic and unstable: being spoken is a sufficient condition for them to change and evolve continually.

3.4 Draft of a Deontology

It is quite surprising that no model, so far, has provided a satisfying description of grammaticalization. We have tried to highlight the intrinsic limitations of the main lines of modeling, to understand why they all seem to fail to account for grammaticalization. From these observations, a few lessons should be learned. We will try to summarize them, sketching thus the draft of a possible deontology for modeling language change.

---

[3]With a few exceptions: the most notable one, the category game, yields a glassy behavior which blocks the language in an ever though slowly changing, non-stable state (Mukherjee et al. 2011); the Fagard and Omodei model (Omodei and Fagard 2013) implements an ad hoc mechanism to ensure that language always changes (a form which is well established comes to carry new meanings); another model (Nadal and Pierrehumbert, unpublished) leads to a state where language yields an unchanging hierarchical structure, but where words constantly exchange their respective roles within this hierarchy.

3.4.1 Role of Analogy

We will first assume the following principle: language is a phenomenon which has no equivalent in Nature; consequently, language evolution is ruled by specific mechanisms and cannot be considered as a subtype of a more general cultural or biological evolution (Croft 2013). The corollary is that analogy should not be used without great care.

Researchers from 'hard sciences' *are* expected to work with analogies, however: otherwise, their specific competences, tools and methods would be of no use. Analogies are useful in the sense that they guide the scientist's attempts to formalize observed features of a given phenomenon. For instance, if it is observed that, when speaking, a human being chooses a given way to express himself among several possibilities, and that this choice is not fully conscious but subject to infinitesimal mental variations, impossible to be accounted for in detail, then one can formalize this observation by seeing it as a stochastic process of decision, a well-studied framework already used to describe many other phenomena.

On the other hand, more systematic analogies (of the scale of the studied linguistic phenomenon as a whole) should be avoided. Analogies should be limited to individual features of a linguistic phenomenon; using them to a greater extent would be fallacious.

3.4.2 Interpretation of Model Mechanisms

The use of analogy can be considered reasonable if any element of the model, any mechanism at work, is susceptible of a clear, unambiguous linguistic interpretation. Thus, when one tries to model a linguistic phenomenon characterized by some specific and well-known quantitative features, it is not sufficient to be able to reproduce them with the model: it is also of crucial importance to know which linguistic realities all the mechanisms which have given rise to this result actually correspond to. This claim may seem blatant; yet, we previously saw that the mechanisms at work in some models may not correspond to any plausible linguistic reality, such as the nonverbal sharing of the intended meaning in naming games. The idea of "language fitness," shared by many population models, also has no clear equivalent in reality.

Furthermore, the mechanisms used in quantitative models are mathematical abstractions. The delicate translation from linguistic mechanisms to their mathematical formulation requires a formalization effort which cannot be the work of 'hard' scientists alone. Very often, hard scientists try to guess from scratch which quantitative mechanisms could be the right ones, while the modeling should proceed along the following steps: the identification, by linguists (possibly led by the insight of the hard scientists who may have an idea of what to look for), of possible linguistic mechanisms, and then, an attempt of formalization to give a mathematical shape to these mechanisms. We give an example of such a way of proceeding in the last section of this work.

3.4.3 Freedom from Hypotheses

A model cannot be free from hypotheses. The formalization effort, the general form of the model, the identified mechanisms, all result from unavoidable working hypotheses. Yet, a model should allow for the testing of these hypotheses, discriminating between them, even if they happen to differ a lot.

Thus, the formalizing effort on which lies the model should not be too heavily biased toward one specific view of language. Building a quantitative model of language change from one specific linguistic hypothesis does not allow the questioning of this very same hypothesis, nor does it favor the opening of a dialogue with other, concurrent views of language. A good formalization should therefore not tell what language is, but it should be able to express in a single, unified mathematical framework, different possibilities of what language might be.

3.4.4 Stylized Facts

The gathering and use of facts is no less important. Indeed, pretending to submit to a scientific deontology implies to make use of observations: ideas, however right they may seem to be, are but speculations as long as they are not grounded on observed, communicable linguistic facts, which are the exact counterpart of the *stylized facts* in complex systems sciences.

Concerning language, these facts can hardly be anything else than corpora-based observations (Biber et al. 1994). But what can these observations consist of? Corpus-based studies help the researcher find occurrences, and observe their context of use (Fagard and Combettes 2013; Ogura and Wang 1996). Finding occurrences is the necessary preliminary of the statistical characterization of a phenomenon—this characterization being the sole feature susceptible of a complete modeling. Looking at the context of occurrences is both a linguistic verification which can in no case be eluded, and the only tool we dispose of to try to understand why and how such a phenomenon has taken place in the past history of language. Without this study of occurrences in context, it seems highly dubious to induce the details of the linguistic mechanism at work. In this very sense, corpus-based studies are an unavoidable premise of the formalization which precedes the modeling attempt itself.

Besides, a linguistic study is often full of ambiguities, uncertainties, or ideas which lack a concrete formulation—where by 'concrete,' we mean leading to a possible numerical implementation. A model cannot be built without filling in those gaps in the understanding of the phenomenon. Hard scientists may have ideas on how to fill in these gaps—using analogies, as we have suggested in Sect. 3.1.

These elements of mathematical nature must have their linguistic counterpart—as we have stressed in Sect. 3.2. And so, once a linguistic interpretation has been proposed, it gives a new insight into the phenomenon. Yet, this insight has to be legitimated by corresponding facts; in other terms, one has to find the traces of its corresponding features in corpora.

Models are of little use if they aim only to describe a given, circumscribed phenomenon. Let us imagine for instance that a viable quantitative description has been found to successfully describe a phenomenon. It would be of course all the more interesting if we manage to extend the original use of the model, in order to find out whether a plausible variation in its core mechanisms could lead to the description of new phenomena. If all the elements of the model have been given a correct, unambiguous linguistic interpretation, then a different result would be more than a mathematical curiosity: it would also correspond to a new plausible linguistic phenomenon, which may be thus identified. It would then be possible to track down the recognized statistical features of this phenomenon in corpora to attest its reality. Once again, corpora appear to be the only reservoir of linguistic facts on which to base any scientific understanding of language change.

Corpora are not, however, the only source of facts. Indeed, they leave apart an important aspect of language: cognitive activity—and yet, one can argue that any language activity is cognitive in its very nature. Corpora give us a glimpse of the *product* of language; but we have to access, one way or another, what *produces* the language: the human mind. Studies in experimental psychology (Scott-Phillips and Kirby 2010) provide us material to explore the cognitive mechanisms at work in language production and, more generally, in human communication. However, while the raw, unprocessed material found in corpora is free from hypotheses, psychological facts on language are the result of experiments, themselves led according to principles, hypotheses, and models.

3.5 A Formalization Attempt: The Grammatheme Abstraction

To illustrate how such principles could take place in an actual case study, we will now present an attempt of formalization and modeling we have developed on the issue of grammaticalization. Details of the model will be given in a future work (Feltgen et al. in preparation).

3.5.1 Linguistic Evidences of the Change

Benjamin Fagard and Bernard Combettes (Fagard and Combettes 2013) have led a statistical, extensive study of a specific case of grammaticalization in French, the partial replacement of *en* "in(to)" by *dans* "in(to)" as one of the main locative prepositions. This statistical study, performed on the French corpus Frantext (ATILF-CNRS and Université de Lorraine 2014), has highlighted several stylized facts:

- the replacement occurs according to a sigmoid-shaped curve (Kroch 1989);
- the timescale of the change is about 50 years;
- the change is actualized in the prose of different authors of that time.

They have, furthermore, tracked this change in a few morpho-syntactic and semantic contexts of use of the locative proposition. It appears that:

- the change within each specific context also follows a sigmoid shaped curve;
- the timescale of the change within each specific context is shorter than the timescale of the overall change;
- the change within each specific context may end before the overall change has ended;
- the change within each specific context may start after the overall change has started.

These facts suggest the following hypothesis: the replacement of en by *dans* occurs in each context separately with the same mechanism; the overall change is the outcome of a spreading of the new gram, *dans*, in more and more contexts of use of the locative preposition. Such a hypothesis has to be more thoroughly tested by supplementary analyses of this change, preferentially on various corpora.

3.5.2 A Formalization Attempt: The Grammatheme Abstraction

We assume that the above stylized facts are generic of the grammaticalization phenomenon. In order to account for them we propose a theoretical framework, which is quite general, and amenable to further generalizations so as to possibly account for most cases of grammaticalization. The goal of this model is twofold: first, to reproduce qualitatively and, as much as possible, quantitatively, the observed facts; second, to shed some light on the underlying mechanisms, and highlight possible links between different phenomena.

3.5.2.1 Zetemes

The basis of our framework is a new formal representation of language, largely inspired from the Utterance Selection Model (Baxter et al. 2006). We first establish a clear distinction between lexical and grammatical semes (for instance, negation could be a grammatical seme, though quite large and unspecified). It does not imply, however, that words themselves are purely lexical or purely grammatical: they may carry semes of both natures. This sharp distinction relies on the distinction

Fig. 3.1 A specific set $O(S, C)$. Each zeteme is the conjunction of a grammatical seme (*upper left color*) and its context of use (*upper right color*). Not all possible conjunctions have to be accounted for. In this example, the set of grammatical semes S is made of three different semes, while the set of contexts C is made of seven different contexts. The borders between different semes have been materialized by a *blue dotted line*

between conceptual and procedural information in relevance theory (Sperber and Wilson 1996), a point more precisely argued in Nicolle (1998). Furthermore, since, as pointed out by Ellis (2008): «Words mean things in the context of other words», we consider that grammatical semes are meaningful only in the context of their use (and may thus convey slightly different meanings in different contexts of use). By 'context of use', we mean something very general, to be understood as the conjunction of a phonetic context, a morpho-syntactic context, a semantic context, and even a pragmatic context. We will thus consider only such a conjunction of a seme and a context of use of this seme.

From a set of grammatical semes S, and a set of contexts C (both chosen according to criteria developed below), we define a more general set $O(S, C)$, encoding all possible conjunctions[4] between the grammatical semes of S and the contexts C (see Fig. 3.1), such conjunctions being called zetemes[5] (informative units). The set $O(S, C)$ is a mere collection of sites, each of them associated to a zeteme (seme-context pairing). It plays a role similar to that of a «meaning space» (Kirby 2001), and provides a framework to represent the linguistic phenomena of interest.

[4]All pairings are not relevant. Some of them may be excluded when representing a situation of interest.

[5]From the Greek root "zêtê" meaning "information".

Since these mathematical entities provide a tool of representation rather than a concept whose every detail would have a unique counterpart in reality susceptible to be tracked and found, we can be quite loose with the specific definition of context; pragmatically, contexts have to be wide and imprecise enough to cover a satisfying number of utterances, and, at the same time, specific enough to offer a rich and accurate description of the situation of interest. It may be considered that the previous remark is also applicable (perhaps with greater caution) to the concept of seme.

Once the degree of accuracy in the definition of semes and contexts has been chosen, the borders of the sets have to be specified. Indeed, including all possibly imaginable contexts and semes from all times and all languages (postulating thus the existence of a universal set of zetemes) is a vain ambition, since it would lead to an unclear mathematical specification of the problem. If one wants to represent a specific situation, then one has to choose accordingly the semes and contexts pertaining to this situation. But we need not deal with these questions when exploring generic and theoretical mechanisms without any reference to a particular, historical, linguistic event.

3.5.2.2 Grammatheme

This framework will help us represent, in a structured and organized way, the past linguistic experience of an individual. This representation will be named a grammatheme, in reference to the thematic organization of the Byzantine Empire.[6]

For an individual, different zetemes may be linked to each other, these links expressing a conceptual proximity between the two sites in the mind of the individual. For a given seme, all contexts of use are likely to be linked, but not necessarily; different semes, if used in similar contexts, could also be linked to each other (see the schematic illustration on Fig. 3.2). Those links are furthermore weighted according to the conceptual distance between the two sites they link up (the weight of the link increases as the two sites are conceptually closer). The grammatheme is thus a network, encoding the conceptual links an individual establishes between different grammatical semes, in different contexts of use.

What actually encodes an individual's grammar (in the sense of (Bybee 2006): «grammar is the cognitive organization of one's experience with language») are the past occurrences of the linguistic elements having expressed a given grammatical seme in a given context of use (i.e., having expressed a given zeteme), linguistic

---

[6]As the *themata* were both the territorial districts of the Empire, and the armies occupying them, the grammatheme deals both with zetemes (the 'territory' of grammar), and with occurrences 'populating' the associated sites; as the number of the *themata* and the borders of the Empire were fluctuating, depending on whether the Byzantine army was keeping them or not, the condition of being expressed determine the belonging of a given zeteme to the grammatheme, the 'borders' of the latter being subject to change.

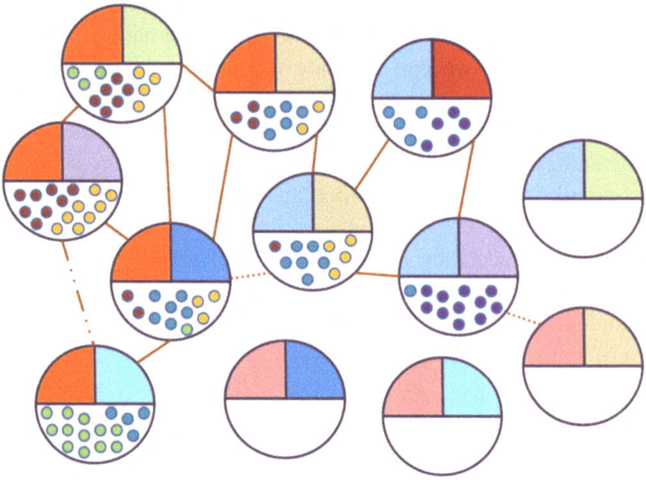

Fig. 3.2 Schematic view of a grammatheme. Zetemes are represented in the same way as in Fig. 3.1. Conceptual links between different zetemes are represented by brown lines. *Dotted lines* are links about to be created. Half plain, *half dotted lines* are links about to be deleted. Each occurrence of a gram is represented by a small colored circle. Different colors stand for different grams. Each time an individual hears or produces an utterance implying a given zeteme, the occurrence of the gram expressing this zeteme is recorded in the corresponding site of this individual's grammatheme

elements we will refer to as *grams*, following (Bybee and Dahl 1989). We describe this past linguistic history of the individual by populating each site of the grammatheme with occurrences of the grams (see Fig. 3.2), used to convey the corresponding zeteme. Within a given site, each gram is associated with as many occurrences as it has been heard and produced in the past history of the individual.

We can formulate the hypothesis, grounded on biological plausibility, that the memory size for each site is actually finite. Therefore, once the memory is saturated, recording a new occurrence implies the deletion of a past memory—i.e., of a past recording of an occurrence. Several ways of implementing this deletion are possible: erasing the oldest recorded memory, or randomly choosing a memory to be erased—the probability distribution function governing this choice being possibly sensitive to the age of the memory.

Not all sites of the grammatheme of an individual are populated by occurrences. The fact that a site is populated reflects his knowledge of the associated zeteme, or at least the (possibly subconscious) awareness of the presence of this pairing in his linguistic environment. Thus, some zetemes are of no relevance to understand and describe his grammatheme. Yet, some unpopulated sites can be related to populated sites by a conceptual link, and so become populated. Thus, an individual may happen to acquire a new grammatical seme (Carlier 2007) which he was unaware of in the past, or to use a seme in a new context.

When a majority of recorded occurrences within a site are occurrences of the same gram, we consider that this gram dominates the site. We can then identify the function or meaning of a given gram—in the structuralist sense—with the dominion of this gram in the individual grammatheme.

Finally, the conceptual links between sites are susceptible of change. They can be reweighted, and even added or removed: similarly populated sites (i.e., sites populated by the same grams) will be more likely to be linked, and conversely, if two sites are populated by different grams, the conceptual distance between them can be revised to become larger than it was.

3.5.2.3 The Grammatheme of a Given Language

How should we characterize a human language within this framework? Like species in biological evolution, languages do not exist as a concrete, directly observable entity. Yet, the ability of individuals to viably reproduce with each other, and the sharing of common traits, allows a regrouping of these individuals into species. Quite similarly, one can regroup idiolects into languages (with the crucial difference that an individual can speak several languages): languages, like species, may be seen as 'existing abstractions;' yet it is not easy to characterize them.

A simple, yet limited way of doing this is by averaging over the idiolects of the individuals recognized to speak the same language, and consider this average to be a satisfying picture of this language. There is an obvious flaw in this characterization: it presupposes that we can identify the sharing of a common language, before characterizing it. It is thus theoretically inadequate. However, it is practically of some relevance, since the identification of a common language is relatively straightforward: interintelligibility can for instance serve as a first, rough criterion to consider that two individuals speak the same language.

Despite these difficulties, we assume this approximation to be acceptable. Indeed, corpus-based studies rarely take into account regional or community differences, and assume the language to define an entity susceptible of being observed. Thus, we can consider the grammatheme of a given language as the grammatheme of a hypothetic, prototypical individual, changing as he 'speaks to itself.' Indeed, we may consider that this prototypical individual produces occurrences on the basis of his past history, and records these same occurrences he has produced. This approximation is in the spirit of the "mean field approximation" in physics and the "representative agent" description in economics—with the same types of qualities and drawbacks.

It should be kept in mind that the mechanisms of producing and recording occurrences (and of redefining the conceptual links of the grammatheme network) at work in the prototypical individual, may be different from the mechanisms characterizing a regular individual. Only a careful statistical derivation could give a satisfying idea of how these self-changing mechanisms implemented in the prototypical individual can be built from and depend on actual, realistic communicative mechanisms between the individuals. For now, the grammatheme leaves this important matter as an open question.

In the following, we will consider a representative agent, embodied with minimal communicative mechanisms we believe to be of relevance for an individual: the ability to produce utterances, to understand them correctly, a desire to be expressive, and the possibility to conventionalize an invited implicature.

3.5.3 *Modeling Grammaticalization: Testing Hypotheses Within the Grammatheme Framework*

We try here to answer the following question: how a gram, so far absent from a grammatical paradigm, and entering it at a given time through a given context of use, becomes fixed (in the same sense of fixed as in population genetics) in this paradigm. To this purpose, we will consider a scheme very similar to the utterance selection model (Baxter et al. 2006). The prototypical individual tries to express a given grammatical seme in a context of use (thus corresponding to one zeteme). To do this, he has to choose a gram, and to produce an occurrence of this gram, occurrence which will be subsequently recorded in the memory of this site. The choice itself is driven by the specific memory (i.e., the set of recorded occurrences) associated with the site, thus setting up a Markovian process for the production of occurrences (a process in which, at each time, the next stage depends on the state of a given object at that time; in this case, memory). The future of a gram is thus determined by this stochastic, Markovian process.

3.5.3.1 Criteria of Choice

We consider that, when the prototypical individual tries to express a given zeteme, it chooses a gram according to two conflicting considerations: it may want to reduce both its cognitive effort and the cognitive effort of its interlocutors, using a word which is as frequent as possible (Diessel 2007); or it may try to use a gram which may be less frequent, but with greater expressivity.

Expressivity is not a property of the grammatheme, but of the grams themselves. Obviously, expressivity should vary as the gram becomes more frequent (through phonetic erosion and semantic bleaching, for instance). We assume that the time-scale governing the change for the expressivity of grams is longer than the time-scale associated to the fixation event within the context of interest. Thus, expressivity can be considered constant in the study of this fixation.

The frequency of grams is a property of the memories encoded in the grammatheme. To compute the actual frequency of a gram in a site, we sum up all the occurrences of this gram in this context, with all the occurrences of this gram in neighboring sites, weighted by the conceptual proximity between sites, and we divide the result by the sum of all occurrences in this site and in neighboring sites, taking into account their respective weights.

3.5.3.2 Origin of the Grams

Where does a new gram come from? We may suggest that some individuals of the population suggest (most often inadvertently) a new way to express a given zeteme, to attract the attention of their interlocutors, to highlight the content of what they have to say, or to fix an insufficiency in the conventional way of expressing it. In the prototypical individual's grammatheme, these individual attempts will be represented by the recording of an occurrence of this new gram.

Another origin of the grams can be proposed, following (Nicolle 1998). According to him, *all* grammaticalization processes start with the conventionalization of an implicature. We can translate this hypothesis into the grammatheme framework, by considering that the conventionalization of an implicature is the appearance of a unilateral conceptual link between one of the grammatheme sites and an external site, associated with a lexical seme. Due to this new link, occurrences recorded in the external site will be taken into account when computing the actual frequency of the grams in the linked up grammatheme site. The external

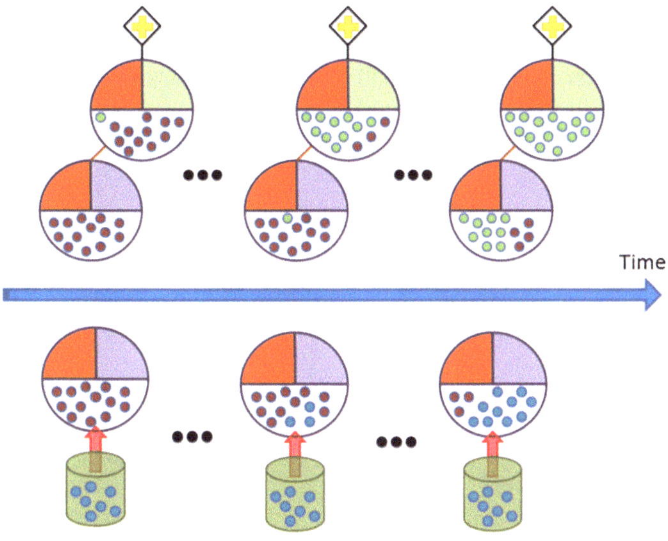

Fig. 3.3 Two scenarios for the fixation of a new gram. Zetemes, grams and conceptual links are represented as in Fig. 3.2. In the first scenario (first row), one context of use is sensitive to expressivity considerations (*green* context, with sensitivity represented by a *yellow* plus sign), while the other is not (*purple* context). The *red* gram is frequent, yet unexpressive, and the green one is newly arrived in the paradigm, but very expressive. In the green context, this new, expressive gram will be favored over the *red* one; in the *purple* context, it will not. Yet, once the *green* gram has dominated the site with the green context, it is able to invade the other site, and finally prevail. In the second scenario (second row), an implicature is conventionalized (*red arrow*). A new gram (the *blue* one) will now arrive from outside the grammatheme (*green* cylinder) to the site of interest, already populated by a gram (the *red* one). If the implicature is strong enough, the *blue* gram will eventually prevail over the red one

site being populated by occurrences of a new gram (the one responsible for the implicature), this new gram will thus be able to 'invade' the grammatheme.

3.5.3.3 Hypotheses and Results

To answer our question (how does a new gram become fixed?), we have considered (Feltgen et al. in preparation) a given site of the grammatheme, where a new gram is supposed to appear. We tested the two following hypotheses, sketched in Fig. 3.3: 1—new grams are able to become fixed because they have a greater expressivity; 2—new grams are able to become fixed because they are the result of the conventionalization of an implicature.

Hypothesis (1) fails as it is: indeed, a new gram never becomes fixed, unless the criterion of choice is dominated by expressivity considerations. This seems highly unrealistic: the repetition of past occurrences seems to be the dominant mechanism in actual communication; otherwise, there would be little understanding between individuals only caring to endlessly innovate. Yet, this hypothesis may be saved by considering that the criterion of choice depends on contexts of use: in some very specific contexts, it is plausible that expressivity considerations are favored over repetition. Those (highly hypothetic) contexts would then make the fixation of a new gram possible, and this gram will then be able (though this next step is in no way automatic) to spread over the rest of the grammatheme (see Fig. 3.4a).

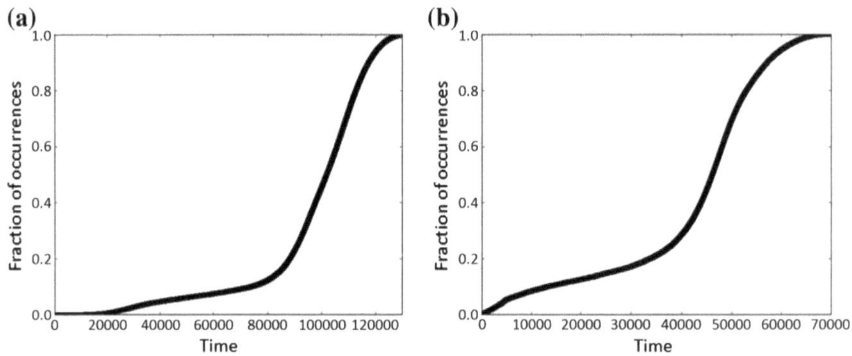

Fig. 3.4 Time evolution of the fraction of occurrences of the new gram (**a**) in the frequency-oriented zeteme (*orange-purple* zeteme in Fig. 3.3), following the first scenario and (**b**) in the entering zeteme, following the second scenario. Time is given by numerical time, i.e., the number of iterations of the model. In the first scenario (**a**), the new gram cannot enter the zeteme at first (0–2000), since grams are chosen to be uttered according to their frequency. Later (2000–8000), as the new gram dominates the expressivity-oriented zeteme, it starts to be used in the frequency-oriented zeteme. The new gram will eventually prevail in both zetemes. In the second scenario (2), the new gram is immediately used, and replaces the older gram according to a sigmoid-shaped curve

Hypothesis (2) gives promising results (see Fig. 3.4b). New grams may or may not become fixed in their entering site, depending on the degree of conventionalization of the implicature (formalized as the weight of the link between the external site and the grammatheme site). This is in line with the claim that conventionalization of implicature is indeed the starting point of any grammaticalization (Nicolle 1998).

3.6 Conclusion

We have proposed a new framework, the grammatheme, which offers a convenient mathematical representation of various hypotheses and linguistic concepts. The model can be developed in many directions. Future works will explore how to implement other important features of grammaticalization, such as phonetic erosion and semantic bleaching, but also further pragmatic considerations A first example in this direction can be found in (Feltgen et al. 2016). It would also be interesting to incorporate some information theory elements, a path already explored in (Fortuny and Corominas-Murtra 2013).

Yet, for now, it is already an accomplishment to be able to compare different scenarios, relying on different hypotheses, of the very beginning of the grammaticalization process. This led us to point out the importance of conventionalization of implicatures. Linguistically, it is far from being new, and (Bybee 2006) provides a comprehensive study on this matter. However, our attempt is one of the first to account for the viability of this process in a quantitative fashion.

In this task, we believe to have fulfilled what is expected from a modeling approach. A modeling approach cannot definitely rule out a possible explanation—one can always think of a mathematical variant, or add a new mechanism which would change everything. But a model can affirm that an explanation actually works—it can attest the efficiency of this explanation. We do not rule out the importance of expressivity considerations: we just prove that conventionalization of implicatures is an effective mechanism, susceptible of implementation, and that it yields the results it had been elsewhere claimed it would be able to.

And this leads to new questions. We have assumed that the conventionalization just happens, without exploring how, or why. We did not even raise the question of what could cause the weight of the conceptual link symbolizing this conventionalization to vary. To answer this question, we could extend the grammatheme representation to the whole language, accounting for purely lexical aspects of language, as well as the important distinction between the two poles of the lexico-grammatical continuum. We would also need to carefully check that they do differ, that they are not blurred into each other in an inaccurate representation. Our approach thus offers an opening towards the investigation of new questions, that it has helped to identify. We therefore, consider the grammatheme as a promising modeling tool to explore the pitfall of grammaticalization.

References

ATILF-CNRS, Université de Lorraine. (2014). Base textuelle FRANTEXT [WWW Document]. URL http://www.frantext.fr.

Baronchelli, A., Felici, M., Loreto, V., Caglioti, E., Steels, L. (2006). Sharp transition towards shared vocabularies in multi-agent systems. *Journal of Statistical Mechanics: Theory and Experiment, 2006*, P06014.

Baronchelli, A., Loreto, V., & Steels, L. (2008). In-depth analysis of the naming game dynamics: The homogeneous mixing case. *International Journal of Modern Physics C, 19*, 785–812.

Baxter, G. J., Blythe, R. A., Croft, W., & McKane, A. J. (2006). Utterance selection model of language change. *Physical Review E, 73*, 046118.

Biber, D., Conrad, S., & Reppen, R. (1994). Corpus-based approaches to issues in applied linguistics. *Applied Linguistics, 15*, 169–189.

Bybee, J. L. (2003). *Phonology and language use*. Cambridge: Cambridge University Press.

Bybee, J. L. (2006). From usage to grammar: The mind's response to repetition. *Language, 82*, 711–733.

Bybee, J. L., & Dahl, O. (1989). The creation of tense and aspect systems in the languages of the world. *Studies in Language, 13*, 51–103.

Carlier, A. (2007). From preposition to article: The grammaticalization of the French partitive. *Studies in Language, 31*, 1–49.

Castellano, C., Fortunato, S., & Loreto, V. (2009). Statistical physics of social dynamics. *Reviews of Modern Physics, 81*, 591–646.

Castelló, X., Baronchelli, A., & Loreto, V. (2009). Consensus and ordering in language dynamics. *European Physical Journal B: Condensed Matter and Complex Systems, 71*, 557–564.

Chomsky, N. (1981). *Lectures on government and binding*. Dordrecht: Foris Publications.

Christiansen, M. H., & Kirby, S. (2003). Language evolution: Consensus and controversies. *Trends in Cognitive Sciences, 7*, 300–307.

Croft, W. (2013). Evolution: Language use and the evolution of languages. In P.-M. Binder & K. Smith (Eds.), *The language phenomenon, The Frontiers Collection* (pp. 93–120). Berlin: Springer.

Cucker, F., Smale, S., & Zhou, D.-X. (2004). Modeling language evolution. *Foundations of Computational Mathematics, 4*, 315–343.

Diessel, H. (2007). Frequency effects in language acquisition, language use, and diachronic change. *New Ideas Psychology Modern Approaches to Language, 25*, 108–127.

Ellis, N. C. (2008). The periphery and the heart of language. In S. Granger & F. Meunier (Eds.), *Phraseology: An interdisciplinary perspective* (pp. 1–13).

Fagard, B., & Combettes, B. (2013). De en à dans, un simple remplacement? Une étude diachronique. *Langue Française, 178*, 93.

Feltgen, Q., Fagard, B., & Nadal, J.-P. (2016). Représentation du langage et modèles d'évolution linguistique: la grammaticalisation comme perspective. *TAL, 55*(3), 47–71.

Feltgen, Q., Fagard, B., & Nadal, J.-P. (in preparation).

Fitch, W. T. (2005). The evolution of language: A comparative review. *Biology and Philosophy, 20*, 193–203.

Fortuny, J., & Corominas-Murtra, B. (2013). On the origin of ambiguity in efficient communication. *Journal of Logic, Language and Information, 22*, 249–267.

Gómez, J.-C. (2007). Pointing behaviors in apes and human infants: A balanced interpretation. *Child Development, 78*, 729–734.

Heine, B. (2003). Grammaticalization. In Joseph, B. D. & Janda, R. D. (Eds.), *The handbook of historical linguistics* (pp. 573–601). Oxford: Blackwell Publishing Ltd.

Hopper, P. J., & Traugott, E. C. (2003). *Grammaticalization*. Cambridge: Cambridge University Press.

John, T., & Bennett, A. (1982). Language as a self-organizing system. *Cybernetics and System, 13*, 201–212.

Keller, R. (1989). Invisible-hand theory and language evolution. *Lingua, 77*, 113–127.

Keller, R. (2005). *On language change: The invisible hand in language*. New York: Routledge.
Kirby, S. (2001). Spontaneous evolution of linguistic structure-an iterated learning model of the emergence of regularity and irregularity. *IEEE Transactions on Evolutionary Computation, 5*, 102–110.
Kirby, S., & Hurford, J. R. (2002). The Emergence of Linguistic Structure: An Overview of the Iterated Learning Model. In A. C. Laurea & D. P. L. Laurea (Eds.), *Simulating the Evolution of Language* (pp. 121–147). London: Springer.
Komarova, N. L., & Jameson, K. A. (2008). Population heterogeneity and color stimulus heterogeneity in agent-based color categorization. *Journal of Theoretical Biology, 253*, 680–700.
Kroch, A. S. (1989). Reflexes of grammar in patterns of language change. *Language Variation and Change, 1*, 199–244.
Leavens, D. A., Hopkins, W. D., & Bard, K. A. (1996). Indexical and referential pointing in chimpanzees (Pan troglodytes). *Journal of Comparative Psychology, 110*, 346–353.
Loreto, V., Baronchelli, A., Mukherjee, A., Puglisi, A., & Tria, F. (2011). Statistical physics of language dynamics. *Journal of Statistical Mechanics: Theory and Experiment, 2011*, P04006.
Meillet, A. (1912). *L'évolution des formes grammaticales*, Zanichelli.
Mukherjee, A., Tria, F., Baronchelli, A., Puglisi, A., Loreto, V. (2011). Aging in language dynamics. *PLoS ONE, 6*, e16677.
Nicolle, S. (1998). A relevance theory perspective on grammaticalization. *Cognitive Linguistics, 9*, 1–36.
Niyogi, P. (2009). *The computational nature of language learning and evolution*. Cambridge, Massachusetts: MIT Press.
Niyogi, P., & Berwick, R. C. (1996). A language learning model for finite parameter spaces. *Cognition, Compositional Language Acquisition, 61*, 161–193.
Nowak, M. A., Komarova, N. L., & Niyogi, P. (2001). Evolution of universal grammar. *Science, 291*, 114–118.
Nowak, M. A., & Krakauer, D. C. (1999). The evolution of language. *Proceedings of National Academy of Sciences, 96*, 8028–8033.
Ogura, M., & Wang, W. S.-Y. (1996). Snowball Effect in Lexical Diffusion, *English Historical Linguistics* 1994: *Papers from the 8th International Conference on English Historical Linguistic* (8. ICEHL, Edinburgh, 19–23 September 1994).
Omodei, E., & Fagard, B. (2013). Cases, Prepositions, and In-Betweens: Sketching a Model of Grammatical Evolution. In *European Conference on Complex Systems* (ECCS, Barcelona, 16–20 September 2013).
Nadal, J.-P. & Pierrehumbert, J. B. (unpublished).
Pawlowitsch, C. (2007). Finite populations choose an optimal language. *Journal of Theoretical Biology, 249*, 606–616.
Pawlowitsch, C., Mertikopoulos, P., & Ritt, N. (2011). Neutral stability, drift, and the diversification of languages. *Journal of Theoretical Biology, 287*, 1–12.
Peyraube, A. (2002). L'évolution des structures grammaticales. *Langages, 1*, 46–58.
Pierrehumbert, J. B. (2001), Exemplar dynamics: Word frequency, lenition and contrast. In Bybee, J. L. & Hopper, P. J. (Eds.), *Frequency and the emergence of linguistic structure* (pp. 153–157). Amsterdam: John Benjamins.
Prévost, S. (2003). La grammaticalisation: unidirectionnalité et statut. *Français moderne, 71*(2), 144–166.
Puglisi, A., Baronchelli, A., & Loreto, V. (2008). Cultural route to the emergence of linguistic categories. *Proceedings of National Academy of Sciences, 105*, 7936–7940.
Reali, F., & Griffiths, T. L. (2010). Words as alleles: connecting language evolution with Bayesian learners to models of genetic drift. *Proceedings of the Royal Society of London B: Biological Sciences, 277*, 429–436.
Schulze, C., & Stauffer, D. (2005). Monte Carlo simulation and the rise and fall of languages. *International Journal of Modern Physics C, 16*, 781–787.

Scott-Phillips, T. C., & Kirby, S. (2010). Language evolution in the laboratory. *Trends in Cognitive Sciences, 14*, 411–417.

Sperber, D., & Wilson, D. (1996). *Relevance: Communication and Cognition*, (2 ed.). Oxford ; Cambridge, MA: Wiley-Blackwell.

Steels, L. (1995). A self-organizing spatial vocabulary. *Artificial Life, 2*, 319–332.

Steels, L. (1999). *The talking heads experiment, words and meanings.*

Steels, L. (2011). Modeling the cultural evolution of language. *Physics of Life Reviews, 8*(339–356), 4.

Steels, L., & Belpaeme, T. (2005). Coordinating perceptually grounded categories through language: A case study for colour. *Behavioral and Brain Sciences, 28*, 469–529.

Traugott, E. C., & Dasher, R. B. (2001). *Regularity in semantic change.* Cambridge: Cambridge University Press.

Victorri, B., & Fuchs, C. (1996). *La polysémie - construction dynamique du sens, Langue, raisonnement, calcul.* Paris: Hermès Science Publications.

Vogt, P., & De Boer, B. (2010). Language evolution: Computer models for empirical data. *Adaptive Behavior: Animals, Animats, Software Agents, Robots, Adaptive Systems, 18*, 5–11.

Wedel, A. B. (2006). Exemplar models, evolution and language change. *The Linguistic Review, 23*, 247–274.

Zanette, D. H. (2008). Analytical approaches to bit-string models of language evolution. *International Journal of Modern Physics C, 19*, 569–581.

Chapter 4
The Case for Cognitive Plausibility

Pietro Perconti

4.1 Why Complexity for Language?

Over the last decades, many epistemological frameworks have been used to understand how language works, including structuralism, generative grammar and cognitive science. Is complexity a better framework than others? In what follows, I will not compare these approaches to decide which is—generally speaking—the right one. On the contrary, I will consider complexity in a mere instrumentalist perspective. The rationale of philosophical instrumentalism cannot be covered here, but we can nevertheless appreciate its theoretical advantages, so to speak, in use. The counterpart of philosophical instrumentalism in computational modeling can be grasped by the following Box and Draper (1987) claim: «All models are wrong, but some are useful».

The kind of observations which the need of complexity arises from, in fact, seems to suggest the *use* of this account to handle the following issues. Only in the case of success we can consider whether complexity could be proposed as a comprehensive framework to understand language. A first major problem comes from the acquisition of language. The cognitive science, language acquisition has been almost a theoretical puzzle because of the amount of information to be acquired and the rapidity of its acquisition. How can a child find the right syntactic, semantic and pragmatic regularities in the kaleidoscope of linguistic inputs she/he is exposed to? This is exactly the basis of the well-known argument of the poverty of the stimulus, so popular in the field of generative grammar. The theory of complexity shows its full potential in modeling the computational difficulties of human language acquisition. Language acquisition, in fact, is a major field of investigation

P. Perconti (✉)
Department of cognitive science, Università di Messina, Messina, Italy
e-mail: pietro.perconti@unime.it

F. La Mantia et al. (eds.), *Language in Complexity*,
Lecture Notes in Morphogenesis, DOI 10.1007/978-3-319-29483-4_4

in the study of complexity of language (Clark and Lappin 2013; Westergaard 2014; Phillips and Pearl 2015).

The semantics of words and sentences, however, is the best context in which complexity theory shows its explanatory power (Barton et al. 1987; Niyogi 2006). According to Noam Chomsky, the compositionality of meaning is (with boundless recursivity, generative power, discrete infinity and constituency) a unique component of human language. The principle of compositionality, i.e. the original Frege's thesis according to which the meaning of a syntactically complex expression is a function of the meanings of its constituents, and the rules in which these constituents are combined, is a key feature of human language. Although the original formulation of the principle by Gottlob Frege was mathematical in kind, compositionality of meaning has never been satisfactorily modeled.

How can the mere syntactic structure guarantee that an unexpected and complex meaning arises from the basic constituents? Although Wolfram Hinzen's following statements are not formulated for *language* complexity, I suspect that, even this one shows some semantic "jumps" like those he mentions: "Syntactic combinatorial complexity may not merely be a linear progression of a sequence of the same object, but involve 'jumps' in representational complexity, with asymmetric entailments following from the operations recursively applied to low-dimensional objects" (Hinzen 2012, p. 368). Modeling these "jumps" in semantic complexity is a real challenge, but perhaps it is also the only way to pursuit a formal representation of the compositionality of meaning.

Another crucial issue from which the need of complexity arises is based on the commitment to computational psychology. The problem comes from the following circumstance: if you are committed to the representational computational theory of mind, you have to handle the problem of computational tractability. Using a given amount of time and space to solve a cognitive problem, in fact, is like a nightmare for computationalists. As we will consider in some detail in the Sect. 4.3, the case of specification of the intentional content is, for example, a serious problem for computational tractability. But first, in order to get a general view on the problem, take into consideration the following statements by Eric Sven Ristad:

> Human language is produced, comprehended, and acquired. Each of these cognitive activities is a process, a mapping from inputs to outputs. (…) In short, the language user is continually preforming certain computations, called *linguistic computations*, that produce and comprehend linguistic utterances and acquire new languages on exposure to linguistic evidence. The input to a language computation include all information that affect the output of computation. The output of acquisition is a hypothesis about the target language, and the output of comprehension and production is a linguistic representation (Ristad 1993, p. 1).

Are these computations, which the language user has to perform, actually tractable? Furthermore, are they cognitively plausible? Such questions may of course be approached from many points of view, but on the whole they are arguments in favour of a more embodied computational psychology. In the end, what we really need, in order to modeling language complexity, is a new embodied and cognitively plausible computational psychology—easier said than done.

4.2 Tractability

As outlined above, if we are interested in promoting a computational psychology which is also cognitively plausible, then we are committed to the concern of tractability of the cognitive functions, that is, we need to understand what exactly means that a given cognitive function is tractable, or not. According to the Tractable Cognition thesis (Van Rooij 2008), what a given human cognitive capacity can actually do is constrained by the limited resources for computation which are typical of human brains. Arguing for a theory of tractable *competence*, Marcello Frixione (2001) distinguishes between computationally tractable and intractable tasks as follows: "Tasks are usually assumed to be computationally intractable, if they require computational resources (time and space of memory) growing exponentially with respect to the length of the input (…). Tasks are assumed to be computationally tractable, if they require resources that grow at most polynomially with respect to the length of the input" (Frixione 2001, p. 381).

In this perspective, the set of functions describing *actual* cognitive capacities is conceived as a subset of the set of functions describing *possible* cognitive capacities and this latter, in turn, is a subset of the set of functions describing *tractable* functions. Iris van Rooij offers the following picture to represent the scaffolding of functions according the Tractable Cognition thesis (Van Rooij 2008, p. 946).

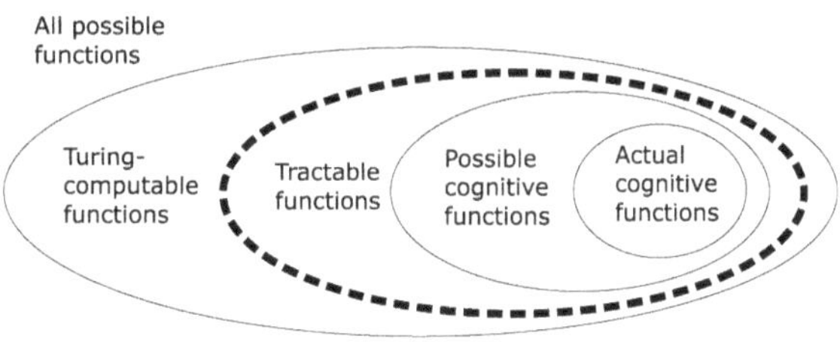

Although one might think otherwise, the constraints of computational tractability express a crucial naturalistic concern. In fact, the limits of what is tractably computable depend on not idealized features of computation, like time, space and memory span. The point is that brains are computational organs, not abstract and idealized machines. They are biological organs characterized by several concrete constraints, like the deterioration of biological functions due to the age, limitations in blood flow and hence in glucose delivery and so on. For this reason, tractability is both a computational and a biological concern. In fact, the problem of tractability is the main pathway which goes from the representational computational theory of mind to naturalistic psychology.

As it is well known, many years ago Jerry Fodor expresses his skepticism about the possibility to keep naturalistic and computational psychology together (Fodor 1974, 1980). But, now—in the age of embedded embodied cognitive science—this agreement seems mandatory. The reasons why naturalistic concerns are nowadays urgent include the social prestige of neuroscience and the influence that biology, in general, has in the new science of mind.

4.3 Cognitive Plausibility

In cognitive science the expression “cognitive plausibility” has too many meanings, and too divergent one from the other. William Kennedy (2009) distinguishes between three meanings in the fields of cognitive modeling, artificial intelligence and social simulation. In cognitive modeling the search for plausibility is aimed to represent single components of cognition, such as memory and vision, and to match the errors human beings make. In artificial intelligence, a given cognitive process is called plausible if the balance between inputs and outputs is comparable to the analogous in the human beings, and if a given computational architecture is successfully modeled. In the social science, the absence of the experimental evidence convinced many scientists to use simulations of the individuals’ and groups’ behaviour. In this view, a social simulation can be called “plausible” if it takes into account the constraints of human beings’ “limited rationality” and if it is able to avoid the typical misunderstandings of the “homo economicus” framework.

Another use of the “cognitive plausibility” is aimed at modeling human plausibility judgments. Connell and Keane (2006) argued for a Plausibility Analysis Model, which is characterized by the attempt to model how a judgment is more or less plausible in relation to a target scenario. As a result, Connell and Keane claim that “a highly plausible scenario is one that fits prior knowledge well”, i.e. that agrees with common sense knowledge of a given community of communication. On their part, Phillips and Pearl (2015) adopt a statistical word segmentation strategy, inspired by the Bayesian approach, to measure the cognitive plausibility of their word acquisition theory.

My use of “cognitive plausibility” would like to stress a condition which computational psychology should satisfy in order to be compatible with the evidence of neuroscience, and biology in general. The required condition says that the psychological (and hence linguistic) computations have to be (1) tractable and (2) fit to the ordinary situations they are encoding. Brains have to do with concrete contexts, that is, boundaries in space and time and the influence of the associative networks embodied in the individual’s past. They are not idealized machines in front of an idealized world, but flesh and blood in search for life and survival. Since it is impossible to include every detail of the modeled target scenario in the computational architectures, a model has to simplify the task.

The crucial problem cognitive plausibility has to deal with is a paradox, i.e. a kind of the “too much or too little information” paradox. On one hand, brains have to take into consideration a lot of information from the context and from their past. As embedded embodied cognitive devices, brains are dependent on a virtually infinite amount of contextual inputs. Representing such an impressive amount of information is a desperate (intractable) challenge for finite brains. On the other hand, if we consider brains as computers which are encoding an unrealistic world in an abstract manner, computations become virtually computable (in an infinite time), but the relation itself between the brain and the world seems implausible. Considering the amount of neglected information, one can wonder whether "understanding the world" is still the right phrase to use.

The paradox of the "too much or too little information" tells us that brains have the problem to encode the *right* amount of information. But, of course, who decides when a given set of information is the right one? The best candidate for this role is the *implicit knowledge*. The main advantage of implicit knowledge, in fact, is that it does not require massive encoding by brains (and computers). It is characterized by a sensorimotor representational format more than a propositional one (Perconti 2013). Using implicit knowledge, a cognitive system behaves in a fast and frugal way. Among others, Haselager et al. (2003) claim that the assumption itself of representations in cognitive science has several disadvantages and consider the dynamical systems theory approach as a non-representational alternative. Gualtiero Piccinini argued that the notion of computation does not presuppose that of representation (Piccinini 2008).

But, if frugal encoding by means of implicit knowledge is a plausible model of the way human brains are environmentally fit thanks to their specific way of computing, this framework could combine the advantages of a cognitive plausible and tractable computation with the advantages of *computational representational* psychology. Taking into account the concern for cognitive plausibility, we can bridge the gap between dynamical and pure computational supporters. According to van Gelder (1998, Sect. 6.10), "dynamical systems can compute, i.e. be computers, without needing to be digital computers". These latters, according to him, compute in a very different way from dynamical systems, because—in the Roberto Cordeschi and Marcello Frixione words—"dynamicist explanations have no mechanistic counterpart" (Cordeschi and Frixione 2007). If implicit knowledge, however, does the hardest job and allows a fast and frugal computation based on a tractable amount of represented information, we are not to choose between the dynamical and the computational advantages.

Thanks to its ability to handle implicit and common sense knowledge, brains are computational devices light enough to be able to ensure tractability. As it is well known, for classical Turing-computable functions common sense knowledge is instead a real puzzle (McCarthy 1959; McCarthy and Hayes 1969). But, brains use many shortcuts to be environmentally and socially fit. They include non-conceptual knowledge (Evans 1982), fast and frugal heuristics (Gigerenzer 2007; Gigerenzer et al. 2011), and implicit encoding of sensorimotor schemata and thoughts (Ito 2012). All this implicit knowledge can be taken from granted and rapidly processed by the brains. «Light» computations by means of implicit knowledge allow the brains to be representational machines focused on the key features of the perceptual and social scene. Common sense and implicit knowledge is both what biological machines are able to handle better and what artificial computers do worst.

The characterization of the intentional content offers a good example of the strategy human brains use to transform, by means of implicit knowledge, a virtually intractable problem in a tractable one. Mindreading is the ability which allows humans to shape, through the vocabulary of the intentional ontology, the social structures we live by—to mention a popular slogan by George Lakoff and Mark Johnson. In order to be able to do this job, however, the mind-reader needs to specify the content of the intentional states. To attribute a mental state to another individual, or to oneself, one must be able to specify its content. However, we cannot take for granted what exactly the phrase "specification of the intentional content" means.

The empirical findings from cognitive ethology should prevent us to consider the specification of the intentional content like a complete procedure. In fact, if we mean the specification of the intentional content as a complete specification, then we have to conclude that animals do not have beliefs at all—an implausible thesis to maintain nowadays, if we take into consideration the empirical evidence coming from the field of cognitive ethology (Perconti and Zeppi 2015). Decades of empirical discoveries in animal observation strongly support an ontological commitment about animal mind, and this in turn is an *a fortiori* argument in favour of the thesis of realism in human intentional ontology. Nevertheless, we have to admit that an accurate description of this *real* intentional content is in point of fact impossible. We have to conclude, therefore, that what humans actually do when they attribute a mental state to another (human or other animal) individual is based on an *incomplete* specification of the intentional content. This is possible just because most of the information which brains process in mind reading is taken for granted thanks to the «light» encoding of implicit knowledge.

On the whole, mutual understanding seems a kind of tentative effort heavily based on implicit knowledge and frugal encoding of pertinent environmental features. It is just due to this modesty strategy in the specification of intentional content that mind reading is a computationally tractable cognitive process and a social practice which works in the everyday situations. And it is due to the analogous modesty strategy used in encoding social and perceptual features that brains are efficient machines, i.e. cognitive plausible computers able to transform virtually intractable problems in a simple routine work.

References

Barton, E.; Berwick, R. C.; Ristad, E. S. (1987). *Computational complexity and natural language. Cambridge* (Mass.): MIT Press.

Box, G., & Draper, N. (1987). *Empirical model building and response surfaces*. New York: John Wiley and Sons.

Clark, A., & Lappin, S. (2013). Complexity in language acquisition. *Topics in Cognitive Science, 5*(1), 89–110.

Connell, L., & Keane, M. T. (2006). A model of plausibility. *Cognitive Science, 30*, 95–120.

Cordeschi, R., & Frixione, M. (2007). Computationalism under attack. In M. Marraffa, M. De Caro, & F. Ferretti (Eds.), *Cartographies of the Mind* (pp. 37–49). Berlin: Springer.

Evans, G. (1982). *The varieties of reference*. Oxford: Clarendon Press.

Fodor, J. (1974). Special sciences (Or: The Disunity of Science as a Working Hypothesis). *Synthese*, Vol. 28, No. 2 (Oct. 1974): 97–115.

Fodor, J. (1980). Methodological solipsism considered as a research strategy in cognitive science. *Behavioral and Brain Sciences, 3*, 63–73.

Frixione, M. (2001). Tractable competence. *Minds and Machines, 11*(3), 379–397.

Gigerenzer, G. (2007). *Gut feelings: The intelligence of the unconscious*. New York: Viking Press.

Gigerenzer, G., Hertwig, R., & Pachur, T. (Eds.). (2011). *Heuristics: The foundations of adaptive behavior*. New York: Oxford University Press.

Haselager, W. F. G., Bongers, R. M., & van Rooij, I. (2003). Cognitive science, representations and dynamical systems theory. In W. Tschacher & J.-P. Dauwalder (Eds.), *The dynamical systems approach to cognition: Concepts and empirical paradigms based on selforganization, embodiment, and coordination Dynamics. Studies of Nonlinear Phenomena in Life Science* (Vol. 10, pp. 229–242). Singapore: World Scientific.

Hinzen, W. (2012). Syntax in the atom. In M. Werning, W. Hinzen & E. Machery (Eds.), *The Oxford handbook of compositionality* (pp. 351–370). Oxford: Oxford University Press.

Ito, M. (2012). *The cerebellum: Brain for an implicit self.* New York: Pearson.

Kennedy, W. G. (2009). Cognitive plausibility in cognitive modeling. *Artificial Intelligence, and Social Simulation*. Poster presented at the International Conference on Cognitive Modeling. Manchester, UK. 24–26 July 2009.

McCarthy, J. (1959). Programs with common sense. *Proceedings of the Teddington Conference on the Mechanization of Thought Processes*, 756–791 (Her Majesty's Stationery Office, London).

McCarthy, J., & Hayes, P. J. (1969). Some philosophical problems from the standpoint of artificial intelligence. *Machine Intelligence, 4*, 463–502.

Niyogi, P. (2006). *The computational nature of language learning and evolution*. Cambridge (Mass): MIT Press.

Perconti, P. (2013). Two kinds of common sense knowledge (and a constraint for machine consciousness design). *International Journal of Machine Consciousness, 5*(1), 95–101.

Perconti, P., & Zeppi, A. (2015). Mindreading, intentional content and computational tractability. *Anthropology and Philosophy. International Multidisciplinary Journal, 11*, 2014/2015: 121–132.

Phillips, L., & Pearl, L. (2015). The utility of cognitive plausibility in language acquisition modeling: Evidence from word segmentation. *Cognitive Science*, pp. 1824–1854.

Piccinini, G. (2008). Computation without representation. *Philosophical Studies, 137*, 2.

Ristad, E. S. (1993). *The language complexity game*. Cambridge (Mass.): MIT Press.

van Gelder, T. J. (1998). Disentangling dynamics, computation, and cognition. *Behavioral and Brain Sciences, 21*, 40–47.

Van Rooij, I. (2008). The tractable cognition thesis. *Cognitive Science, 32*(6), 939–984. doi:10.1080/03640210801897856.

Westergaard, M. (2014). Complexity and conflicting grammars in language acquisition. *Second Language Research, January 2014, 30*, 85–89.

Part II
Complexity, Semiotics and Enunciation Theory

Chapter 5
System and Structure. Semiotics as Encyclopaedic Theory of Complexity

Claudio Paolucci

5.1 Semiotics and Complexity Theory

Semiotics has perhaps missed the complexity train. It missed it when it was an «imperialist discipline» (Cf. Eco 1975: 17), at the heart of the cultural project of many other disciplines who adopted its models and very particular point of view. It certainly missed it later, when this cultural centrality was lost and semiotics—instead of accepting new cultural challenges—opted to restrain itself to a descriptive discipline of «cultural objects» (texts), handled through a series of «heuristic-oriented analysis models».[1] Perhaps—and this is a thesis that I will attempt to argue and revive in this work—the cultural centrality of semiotics was lost precisely because it was unable to meet the challenge of complexity.[2]

Because, on the one hand, the train of complexity was missed when semiotics insisted too much on the idea that its structures defined above all an analysis method,[3] based on the metalinguistic interdefinition of its categories.[4] While, on the other hand, the complexity train was missed when, at the epistemologic level, a weakly phenomenological path was chosen, where the discipline's object was thought to be constructed from the generative project of the discipline itself, «correlate of its theoretical project of description», in the words of Greimas and Courtès.[5] Quite a weak

---

[1]Cf. Fabbri (2005).

[2]There are various reasons for that, not least the total indifference in the theory of systems, epistemology of models and those scientific reflections that, on the contrary, guided Eco in writing books such as *The Open Work* and *A Theory of Semiotics* (but also, later, *Kant and the Platypus*).

[3]Cf. Eco (1968), Greimas and Courtés (1979), Fabbri (2005).

[4]Cf. Greimas and Courtés (1979).

[5]Greimas and Courtés (1979): "Text". I worked extensively on these points in Paolucci (2010), Chap. 2.

C. Paolucci (✉)
Dipartimento di Filosofia e Comunicazione, Università degli studi di Bologna, Bologna, Italy
e-mail: c.paolucci@unibo.it

F. La Mantia et al. (eds.), *Language in Complexity*,
Lecture Notes in Morphogenesis, DOI 10.1007/978-3-319-29483-4_5

pair that of a theory that only produces a metalinguistically interdefined model which, nonetheless, it is thought to be even capable of constructing its own object. One might be tempted to hope for its inversion, perhaps following Hjelmslev's lesson, where theory produces arbitrarily—i.e. independently from experience—a structural object capable of adapting to experience, showing some aspects that were not directly perceivable in the experience of the object itself.[6] Surely the semiotic option that attempted to keep together «interdefined model» and «constructed object» offered a very weak—too weak—proposal at a time that featured a strong comeback of natural sciences and ontology, together with a violent thrust towards the naturalisation of humanities, which semiotics battled with blunt weapons, from a position that was by the time peripheral and rearguard. And, in my opinion, this was a huge missed opportunity.

Because complexity theory would have offered quite a different destiny for the semiotic idea of system, which, as is well known, was the term that Saussure used instead of the better known «structure» as the main theoretical object of semio-linguistic epistemology.

> The system refers to the real rather in depth: it is more real, because more rooted in the physis and connected to it, than the old, almost artificial object in its pseudo-realism; at the same time, it refers rather in depth to the human mind, i.e. the subject, who is also immersed culturally, socially, historically. It requires a physical science that is at the same time a human science [my translation] (Morin 1977: 161).

Always on the frontier between physis and nomos (cf. Coquet 2007; Petitot 1985), always located beyond the opposition between naturalism and culturalism (cf. Paolucci 2012), semiotics could have met the challenge of complexity, which seemed to outline a perfect placement for its theory of system situated beyond, or on this side, of the objectivity of facts and their theoretical representations. And yet none of this happened.

In this work, I will then try to tie again some links together that came undone without a real reason for doing so, in the attempt to demonstrate a connection between complexity theory and semio-linguistic epistemology that I believe is unbreakable.

5.2 The Origin of Semiotic Undertaking

When he was preparing to systematise its theory, Eco (1978) was fully aware that the semiotic endeavour, with its double Peirce-Saussure foundation, was grounded on a «new sensitivity», a «new philosophical Kunstwollen» capable of «producing a sign-oriented culture». Which one?

The so-called «linguistic turn» of the twentieth century gave us a certain distinction or correlation between object and representation on the one hand, and

[6]This is also the definition itself that Peirce gives for iconism, i.e. the categorical principle that in his opinion is at the basis of every cognition.

between fact and theory on the other; between that which is of the order of the facts of objectivity and that which is of the order of their theoretical representation, independently of the perceptive, linguistic, cognitive, sensitive, social, cultural, mental or neuronal nature that is then acknowledged to this same representation. A large part of contemporary philosophy is often defined in function of the means of variously distributing the complex relations between these two fields, to recognise their specificities and mutual interrelations, to draw a series of boundaries that at the same time divide and are common to these two orders, in order to thus define their different connections. At times, the dispersion point where fact and theory are intertwined and joined until they fade in the distance is cited; other times, their distinct, sharp, razor-edged boundary like the line we want to use to trace the difference between the order of objectivity and the order of its representation is underlined. For example, on the one hand, all contemporary hermeneutics is founded on a fundamental difference, called ontological difference, which establishes, distributes and separates an ontic level of objectivity (beings, Seinde) from an ontological regime of light that makes it visible (being, Sein), and that thus defines the conditions of possibility of its representation. In the same way, on the other hand, in an Anglo-Saxon environment, it is the different modulation of complex relations between fact and theory that represents the central theoretical turning point from which a whole series of differentiated positions unfold, including, by way of example, the underdetermination of the theory in relation to the facts in Quine, the constructivism of the theory on the factual datum in Goodman, the paradigmaticity of theory on what is seen in the factual datum in Kuhn, the falsifiability of theory from facts in Popper, the impurity of facts in relation to the theory they are always loaded with in Feyerabend.

This distribution is even more clearly marked in the positions following the linguistic turn, such as for example in those of the so called «new realism», in which they programmatically give a «welcome back» to reality,[7] or in those of many «naturalist» theories.

Now, the «linguistic turn» of semiotics is completely neutral with respect to these distributions. It is foremost defined by the discovery of a third order located beyond, or on this side, of the objectivity of facts and their theoretical representations. This new order of Third is then identified with the discovery of the specificity of a semiotic order, beyond any possible dualism between object and representation, and between fact and theory. The refusal to confuse this strictly semiotic third order with that of the objectivity of facts, with that of their theoretical representation and with the complex relations that can be identified between these two fundamental constituents, is then the fundamental dimension from which we recognise the semiotic paradigm in its double Peircean-structuralist foundation.

Here it is indeed that beyond objects, the representamina and their specific relations of Firstness and Secondness, Peirce identified a third level formed by semiotic magnitudes, the interpretants, that are able to redistribute the relations

[7]Cf. Ferraris (2012), Ferraris and De Caro (2012).

between the elements we believe we can identify within the other two orders. If indeed the representamen is 1 (order of representation) and the object is 2 (order of objectivity), for Peirce the semiotic is then always essentially of the triadic order of Three:

> by «semiosis» I mean…an action, or influence, which is, or involves, a cooperation of three subjects, such as a sign, its object, and its interpretant, this tri-relative influence not being in any way resolvable into actions between pairs. (CP 5.484)

With this idea of semiosis as a structure irreducible to actions between pairs, Peirce thinks of the semiotic as the location of a structural complexity irreducible to any type of binary relation,[8] and therefore to any distribution between the object and its representamina (cf. Paolucci 2010: Sect. 1.5).

Perhaps here everything started from linguistics: beyond the word in the objectivity of its auditory parts, beyond the acoustic images, the concepts and representations associated with the words, the structuralist linguist indeed discovered a completely different element, a «structural object». For example, the phoneme is indeed manifested in letters, syllables and sounds, but without deriving from them nor reducing to them, since it represented their condition of possibility. At the same time distinct from both the auditory (phonic) substances and the acoustic images to which it was associated, the phoneme was embodied by them, but in itself it was only defined by the plane of immanence in which it held differential relations with other phonemes and with heterogeneous elements compared to the plane of immanence taken into account (the phoneme was in fact the smallest element of the plane of the signifier capable of producing differences on the plane of the signified). This plane of immanence therefore held its own reality, a systemic reality that defined the identity of the elements as emerging from the structure of the relations internal to the system.

As is well known, these are two characteristics that constitute the complexity theory. Morin (1993) indeed defines complexity as «a fabric of heterogeneous constituents that are inseparably associated», and called this fabric «system», identifying one of the main characteristics of complexity in an interdependent relation between the elements: «in a complex system, the relations between the elements become more important than the nature of the elements themselves [TN: my translation] (Gandolfi 2008: 41). It is precisely this structural interdependence that gives life to emerging properties that constitute the system:

> when the system exceeds a critical threshold of complexity, structures that involve and organise in a coordinated manner many of the elements of the system are formed. …When a system organises itself, structures and global properties that were previously nonexistent emerge. These new properties become dominant for the overall behaviour of the system; the individual elements are subservient to the new emerging structure. The German physicist Hermann Haken showed that these global characteristics of complex systems cannot be understood in a simple way starting from the behaviour of their elements. This is because in the new level, the crucial aspect of the system's behaviour no longer consists in the nature

---

[8]Cf. Morin (1986).

of the individual elements, but rather in the relations between the elements… In other words: even studying every single detail of the individual elements that make up a system, it is not possible to understand and predict the overall behaviour of the system. [TN: my translation] (Gandolfi 2008: 43–4)

Hence the constitutive anti-reductionism proper to an approach to complexity. Hence, the central idea of structural linguistics that, in order to understand the behaviour of a language, we must study its system, and not the individual elements. Hence Peirce's central idea that in order to study semiotics, we must first of all study a Logic of Relations.

In short, since its foundation, it seems evident that the place for semiotics was the place for complexity, the place of the system. It is not by chance that, as mentioned, in putting the notion at the core of semio-linguistic epistemology, Saussure did not use the term «structure» at all. On the contrary, he used precisely the term «system». But what system did Saussure have in mind when he asserted that language was a system?

5.3 The System as a Heteronomous Entity of Internal and External Dependencies

In that collection of manuscripts published under the name of *Writings in general linguistics*, Ferdinand de Saussure was able to outline which were the constituent peculiarities of «semiological systems» and what differentiated them from other non-semiotic domains.

Unless I am mistaken, objects of study in other areas can be said to have their own existence, or failing this to encapsulate certain positive things or entities…; The science of language appears to be in a different situation. […] It is fundamental to and inherent in the nature of language that from whichever side one tries to tackle language, there can never be found individual entities, that is entities (or quantities) that can be defined in themselves and… have an independent existence. Let us remember indeed that the object in linguistics does not exist, it is not determined in itself (ELG: 42, 8, 71).

But what kind of entity can a nonindividual entity be, keeping in mind that it has no independent existence, that is not determined in its own right and that is not therefore definable through properties?

Saussure (CLG) identified above all a double dimension that for him (and for me) was distinctive of the whole semiotic endeavour as a whole. This double dimension was constitutive of an element that seemed to depend upon it and which represented the «concrete entity» of the new science (CLG: 125-9). It was a truly new discovery, as simple as it was decisive, whose «strange» and «surprising» character Saussure did not stop emphasising (CLG: 130). This concrete entity, while it did not stop circulating in both language planes, in sounds, in concepts, in acoustic images, in words and phrases, was not, however «immediately perceptible». For that Saussure wondered whether it was «truly given»:

> A language thus has this curious and striking feature. It has no immediately perceptible entities. And yet one cannot doubt that they exist, or that the interplay of these units is what constitutes linguistic structure.[9] (CLG: 130)

For Saussure, the concrete entities of language are not immediately perceptible, because they are constantly covered by the objectivity of the linguistic facts in which they are embodied (sounds, meanings, speech acts, etc.) and by the theoretical representations of linguistics that entrap them (concepts, images, propositions, names, adjectives, cases, etc.). And yet, Saussure said, they are not identified with them, just as the Geneva-Paris train of 8:45 pm is not identified with its locomotive, its wagons and personnel (CLG: 132). Indeed, they belong to a different order, which comes third if compared to all these distributions, where the identity of an element is purely differential and is the effect of a local equilibrium through which it is sustained and determined:

> in semiological systems, such as language, where the elements keep one another in a state of equilibrium in accordance with fixed rules, the notion of identity blends with that of value and vice versa. This is why, ultimately, the notions of identity and value merge. That is why in the final analysis the notion of value covers units, concrete entities and realities. (CLG: 134)

This is what the specificity of «semiological systems» consists of, and this is what entities fill their fields. Outside of this identification, there is no semiotic endeavour: there is another system, as Saussure said. But what identity has a value, since the identity of concrete entities of semiotics «merge» with that of «value» and since value is the element of «semiological systems» ?

> To answer this question, we observe foremost that even outside of language all values appear to be supported by this paradoxical principle. They are always composed:
>
> 1) Of a dissimilar thing that can be exchanged for the thing of which the value is to be determined;
> 2) Of similar things that can be compared with the thing of which the value is to be determined.
>
> Both factors are necessary for the existence of a value. (CLG: 140)

This paradoxical principle concerning the elements of the semiological systems has to be radically asserted: values populate that third immanent order that belongs to the semiotic and do so in a constitutive tension that always refers them inside and outside at the same time, in a dialectic between transcendence and immanence that defines a constitutive feature of complexity theory. Indeed, according to Saussure, a value is always defined by two dimensions, both purely differential and relational. The first dimension a value consists of, in fact, resides in its correspondence with the entities outside the system within which it is considered. For example, the value of a five-franc coin is determined by the fact «that it can be exchanged for a fixed quantity of a different thing, e.g. bread» (CLG: 140). The second dimension instead consists of the relations that a value incurs with other values internal to the system

[9]The original french simply speaks of "language".

to which it belongs. For example, «it can be compared with a similar value of the same system, e.g. a one-franc piece, or with coins of another system (a dollar, etc.)» (CLG: 140). These two relational dimensions, one transcendent and the other immanent to the system considered, are both constitutive of the notion of value itself: a value is given only when exchanged with an «outside» and compared with an «inside» one to the system to which it belongs, since only in this manner can element=x become a value.

This is what happens with nonindividual entities such as those of language: they are not determined, but they are determinable and their determination always occurs through a reciprocal determination. This reciprocal determination is in function of a transcendent dimension (exchange with the outside) and an immanent dimension (comparison with the inside) to the element considered. Saussure thus imports the constitutive principle of Leibniz's differential calculus from mathematics into linguistics: hence, the «differential» nature of language units and semiotics.

> As there is no unit (of any order or nature) which reposes on anything other than differences, a unit can only be imaginary; only difference exists. Yet we have no choice but to proceed on the basis of positive units, for without these it would be impossible to bring order to the mass of facts. But it is essential to remember that these units, though, remain nothing more than an unavoidable expedient of our spirit, and nothing more: the simple positioning of a unit implies a decision to leave out the play of relations, and so grant a fleeting separate existence to just one of them (ELG: 80-1).

In other words, value is a purely differential unit composed of two dimensions which are themselves purely differential: it is a relation between relations, internal and external to the system considered.

This strictly semio-linguistic idea of system, where identity depends on the set of relations and this set of relations defines a stock of relationships that are both homogeneous (between entities of the same system) and heterogeneous (between entities of different systems) defines, from its Saussurean foundation, a semio-linguistic system as a *heteronomous entity of internal and external dependencies*, which define an irreducibly complex system in the complexity theory sense. Not only does one of the main characteristics of complexity consist in the «interdependent relation between elements» that gives rise to a «a fabric of heterogeneous constituents that are inseparably associated» called «system» (cf. Morin 1993), but Morin (1993) precisely insists: (i) on the heterogeneous nature of elements of a complex system; (ii) on the fact that this heterogeneity consists not only of the relations internal to the system, but also of relations between the systems. Hence, a system is complex when the relations constitute the identity and behaviour of the elements of the system and when the system is not closed, but rather when it is open to interact with other systems that are heterogeneous to its own (so much so that this characteristic will then give rise to nonlinear behaviours).[10]

An essential characteristic of a complex system consists in its plurisystemic essence, in its being a system made of other systems.

---

[10]Cf. Licata (2008).

> Human beings belong to a social system, which is part of a natural ecosystem, which is part of a solar system, which is part of a galactic system: They in turn consist of cell systems, which consist of molecular systems, which consist of atomic systems [my translation] (Morin 1977: 112).

Summing up, a system is complex when it is plurisystemic, composed in turn of other systems, systems of systems, relations between systems and systems that merge into one another. It is this characteristic that constitutes semiological systems for Saussure, and it is precisely this feature which is constitutive of the idea of value that has been lost in the passage from the original notion of system to the one, which then became famous, of structure. Indeed, it is not by chance that, since Hjelmslev's definition of structure as an «autonomous entity of internal dependencies», it was exactly the transcendent dimension of Saussurean value, which defined the complexity of a semiotic system, that was erased from the idea of a semio-linguistic system. It was precisely this erasure to give birth to the idea of structure as a closed entity composed by homogeneous and intrasystemic dependencies and to prevent a fruitful encounter between semio-linguistic epistemology and complexity theory.

5.4 System and Structure: Hjelmslev's Reductions and Morin's Criticisms

Beginning with Hjelmslev, structuralist semiotic tradition was never able to grasp the differential relationality constitutive of the semiotic value in its transcendent meaning (first dimension) and always ended with confusing it with the reference to a sample-parameter, if not with a referent in the real world. This is why the structuralist tradition decided to eliminate it from its own epistemology, ending by reducing the Saussurean concept of system to a homogeneous and closed structure, based solely on internal and intrasystemic dependencies. The first to operate this misunderstanding and the relative expulsion of transcendent meaning of the Saussurean value from structuralism was doubtless Hjelmslev (EL: 100):

> The comparison with the exchange value limps on an essential point:…the exchange value is defined by the fact that it corresponds to a determined quantity of goods, and this serves to give it a foundation based on natural data, while in linguistics, natural data are completely absent. Economic value is by definition a term with two facets: not only does it have a constant role with respect to concrete units, but by remaining identical it has the role of variable with respect to an established quantity of goods that serves as its referent. Instead, in linguistics, there is nothing that corresponds to the referent. [my translation]

Rastier himself (2004) did not grasp at all the relational differentiality that constituted Saussurean transcendent meaning and, in a radicalisation of Hjelmslev's idea, ended by supporting the wisdom of structuralist tradition by abandoning it exclusively in favour of the second meaning:

> signification consists of the «internal» value, and this is accorded with the decisive abandonment of any reference. Thus the difference between mutton and sheep is linked to their conjoint presence, and therefore to their differential distribution…but not to an a priori difference between their signification that would be based on the difference of their referents. (Rastier 2004: 3)

Now, must this something dissimilar with which a value can be exchanged to be constituted as value necessarily be a referent, a term-parameter or something similar to a gold reserve? Instead, can it not simply be an «outside» with respect to these differential relations that constitute the first meaning, «outside» with which they exchange and translate continuously?

By identifying «identity» and «value» in its double meaning, Saussure was simply saying that, in order to establish the identity of something, this «something» must always be compared with other elements within its system, but it must also be translated with elements belonging to other systems. For instance, if we wish to determine the semiotic identity of Umberto Eco, we must not only compare his positions with those of Peirce, Lotman, Hjelmslev and Greimas, but we must also understand how these positions inflect and translate in a semiotic domain those of other heterogeneous disciplines: sociology, philosophy, linguistics, cognitive sciences, etc. If we want to determine the political identity of the italian party Sinistra, Ecologia e Libertà (Labours, Ecology and Freedom, SEL), we should not just compare their position in the political topology of coalitions, from which we glean that they are opposed to the centre-right, are more to the left of the Partito Democratico (Democratic Party: PD), and more to the right of Rifondazione Comunista (Communist Refoundation Party); but we must also determine which societal values the party Sinistra, Ecologia e Libertà translates within the political domain, i.e. what do they think of civil unions, stem cells, research, economy, safety and of all those values that transcend the political domain which a party translate and represent within the political domain. But are safety and stem cells the referent for Sinistra, Ecologia e Libertà? Are cognitive sciences and anthropology the referent for Umberto Eco, in the same way the animal «cat» is the referent for the word «cat»?

After all, the absurdity of a referential interpretation of the first meaning of the value was already quite clear in Saussure's example. Is bread the referent for money? And is it a natural datum that stays stable, as Hjelmslev would like to have it? Or is it simply an element that belongs to a heterogeneous system with which an immanent element is translated and with which its identity, i.e. its value, is defined (with a five-franc coin you can purchase a certain quantity of bread, which is a lot less today than it used to be in the past)?

With his first meaning of value, i.e. with that «dissimilarity» with which a semiotic value can always be exchanged, Saussure gives us an image of the structure that offers us a system expressed on a thousand plateaus (cf. Deleuze and Guattari 1980). Far from being one plateau—or perhaps in order to be so—the «semiotic» level constitutively refers to a thousand plateaus with which it is reciprocally determined, and from which its identity is derived. The shape of a structure is that of a system in a constant reciprocal determination with other heterogeneous systems from the one considered. It is a system made of other systems, a constitutively trans-dominial complex system where the identity of a level of immanence is constitutively determined by that of a thousand other levels. And it is the most current and fruitful legacy of structuralism: far from being an «autonomous entity of internal dependencies», as Hjelmslev erroneously stated, a structure is a heteronomous entity of internal and external dependencies. The fact that it is a «structure» is not due to the closure of the system, nor the homogeneity of dependencies. On the contrary, the features of a structure are: (i) the reciprocal determination of the elements; (ii) the fact that their identity does not precede the system of relations considered (intrasystemic as well as extrasystemic); (iii) the fact that it is independent of the substances in which it is embodied.

Here, Saussure allows us to go much further than Hjelmslev and the usual idea of «structuralism»: if the idea of structure was indeed used by Hjelmslev in order to separate homogeneous (or uniform) dependencies from heterogeneous dependencies (relations within language are homogeneous, while those gleaned from extralinguistic factors should be considered heterogeneous to the elements of language),[11] the idea of system was used by Saussure in order to describe the semiological systems as constitutively polysystemic, made of both homogeneous and heterogeneous dependencies. Indeed, it is no accident that Hjelmslev explicitly and deliberately rejected Saussure's first meaning of value.[12]

It is therefore Hjelmslev's intrasystemic concept, rather than the Saussurean plurisystemic one, that Morin refers to when he drafts his criticisms on the idea of structure:

> each concept that is only structuralist, i.e. only interested in reducing systemic phenomena and organisational problems in terms of structures, would entail a large dispersal of intelligibility, a distinct loss of phenomenal wealth, a destruction of complexity. Indeed, the idea of structure only includes a conjunction of a few necessary rules that manipulate and combine some basic units. It therefore remains in the dependency on the paradigm of the order (in this case intrasystemic) and of simple objects. [my translation] (Morin 1977: 152)

---

[11]Cf. Zinna (2001).

[12]However, in Hjelmslev, things are much more complex if we consider not only works such as *Prolegomena to a theory of language* or *Linguistic Essays*, but the true masterpieces of structural linguistics such as The Category of Cases, "Structure générale des correlations linguistiques" (General structure of linguistic correlations) or *Resumé of a theory of language*. In these works there is a place within structural linguistics for everything Saussure believed pertained to the first dimension of value, i.e. all nonhomogeneous dependencies (or "non-uniform", as it is often translated). The key concept in this sense is *fragmentation*, which for Hjelmslev flanks that of analysis. Indeed, *analysis* is the «description of an object by the uniform dependencies of other objects on it and on each other» (R, Def. 3). Instead, *fragmentation* is the description opposite to analysis, i.e. the «description of an object by the non-uniform dependence of other objects on it and on each other» (R, Def. 4). According to Hjelmslev, this is exactly what occurs in the participative relations between tensive terms or inside the «analysis by dimensions». «The common term for analysis and fragmentation is *dissection*» (R, Def. 4), i.e. «cutting», «severing!, as in anatomy, where parts and organs of the body are dissected, sectioned and separated in order to study or describe them. I have worked extensively on these issues, with special attention on the relation between participative oppositions, analyses, nonhomogeneous dependencies and the first dimension of value in Saussure in Paolucci (2010), Chap. 1.

I believe we can glimpse a theoretical and epistemological fear in Morin. The concept of structure drawn from structural linguistics is understood as an intrasystemic order that does not allow neither morphological changes nor the interconnection between different heterogeneous systems (hence the final appeal to «simple objects»). However, as we have shown, we cannot agree with Morin on this interpretation, since behind the «linguistic turn» of semiotics, there was no «desire to reduce systemic phenomena and organisational problems in terms of structures». On the contrary, behind the semio-linguistic idea of system, there was more than in nuce a new sensibility that moved towards complexity theory.

It remains clear, though, that Morin's reading was largely motivated by the direction that semiotic epistemology took later, particularly in its generative form, which ascribed itself the legacy of the structuralist linguistic turn, when it was only its constitutive betrayal. As I have shown elsewhere (Paolucci 2010, Chaps. 1 and 2), it was precisely the expulsion of plurisystemic-ness that belongs to the first dimension of Saussure's value that lead semiotic epistemology to adopt an intrasystemic theory made of homogeneous dependencies within a single semio-linguistic system. In rejecting the Saussurean transcendent meaning of value,[13] semiotics in fact ended up identifying, in Greimas' school, (i) structure with the reciprocal determination of the elements within a single system (rejection of the Saussurean translatability with the «outside»); (ii) this single system with the «semantic microuniverse» later called «text»; (iii) the structuring of this semantic microuniverse with the textual «clôture» («no salvation outside the text!»).

This expulsion of complexity from the semiotic structure is perfectly embodied in the semiotic square, which in generative semiotics represents the idea of structure itself, i.e. an «elementary structure of signification» grounded on a «semantic microuniverse», where relations are generated from a basic binary combination, which perfectly embodies that «destruction of complexity» that Morin ascribed to the idea of structure and structural semiotic epistemology.

However, in semiotic tradition, there were essentially four models that sought to determine more accurately what Saussure generically defined as «system»: the tree, the structure, the matrix and the rhizome. The four theoretical objects that «emblematically» represent these different models are Hjelmslev' hierarchy (p: 114), Greimas' semiotic square (1970, 1983), Fontanille and Zilberberg's tensive structure (1998) and Eco's encyclopaedia (1984, 2007). The first two models aim towards simplicity, the latter two aim towards complexity, but it is the rhizomatic encyclopaedia[14] alone that is thought of as a system made of other systems, able to express and make the Saussurean idea of semio-linguistic system operational.

---

[13]Already in *Raison et poétique du sens*, and precisely in reference to the two meanings of Saussurean value, Claude Zilberberg (1983: 17) very correctly noted that the whole of generative semiotics «is constituted towards the adoption of the second principle and the non-theorised abandonment of the first» [TN: my translation].

[14]Cf. Eco (1984), Carta (2012).

5.5 Towards Encyclopaedic Semiotics of Complexity

In order to understand why the encyclopaedia is a system made of other systems, let us start from this enlightening definition by Jean Petitot:

> Let us consider the lexical-semantic organisation of a work such as the Encyclopaedia. «Articles» are grouped in relatively homogeneous «packets» that identify problems with fairly well-defined contours. …These themes indicate «local» issues governed both by local semantic fields, i.e. systems of concepts and specific conceptual opposition, and regional, theoretical and practical, specific and adequate techniques. These local fields are not independent however: they partially cover each other, they «glue» to one another to generate broader extension fields. For example, the local/global opposition partly covers the centric/acentric, continuous/discrete, universal/particular oppositions, some of which gave rise to other articles in the same Encyclopaedia. …The problem with the encyclopaedia as a global object was fairly well formulated by Umberto Eco in the context of structural semantics [in order to] explain the «socialised competence in the liveliness of its contradictions» [my translation]. (Petitot 1979: 429–432)

An encyclopaedia is always a collage of precise zones, i.e. systems that can be described through a system of specific local oppositions, and vague zones, where the systems overlap, are intertwined and the elements see their identity referred to the translation with other heterogeneous systems governed by other systems of specific oppositions. Thus, the «local/global» opposition in Petitot's example. Here it is how both the encyclopaedia as «textual object that aims to reflect the organisation of our knowledge» (Petitot 1979: 430), and the encyclopaedia as «semiotic model of the logic of semiosis» (Eco 1984), are always systems made of other systems.

It is not by chance that Eco (1984) defines the encyclopaedia as the «library of libraries», «the recorded collection of every interpretation», the «great audiovisual archive» of statements.

For this very reason, if structural, dictionarial or textual semiotic theories appear essentially founded on the ideas of closure, coherence and cohesion, through which they try to circumscribe and describe the allocation of meaning, the encyclopaedia instead assumes the boundaries of a tendentially open, contradictory and non-cohesive system. As is well known, Eco insists in several parts of his work:

(i) On the contradictory nature of the encyclopaedia (noncoherence), on its admitting and encouraging at the same time «if p, then q» and «if p, then not q»;
(ii) On the openness of the encyclopaedia, the fact that it has no defined beginning nor end, so much so that at any moment of semiosis it accepts new interpretations which it does not stop recording and which incessantly modify its own structure (precisely in this regard, Eco insists on the impossibility of a global description of the encyclopaedic network, both due to the fact that this description would already have altered it, and the fact that it is absolutely impossible to have a non-short-sighted and external view on it);

(iii) On its constitutive non-cohesion, since the mastery of encyclopaedic contents changes depending on the sociolectal class, culture, individuals, etc. (cf. Paolucci 2010, Chap. 3).

Clearly, these features refer to what has been defined as the «logical openness» of complex systems, i.e. that constitutive characteristic of complexity that makes impossible to resolve the emerging properties of a system within a single model.

> Adopting a single model means to fix variables and interactions, i.e. a finite and fix number of n vertices, whereas in a highly logical open system they continuously and unpredictably emerge and disappear as a consequence of the internal functional organization of the system. (Licata 2012: 14)

Precisely because of this, the encyclopaedia attempts to manage the complexity of sense, which is the object of semiotics, within an open, noncoherent and potentially contradictory model, based on the collage of heterogeneous systems and the coexistence of local structures and distinct models.[15]

This strictly encyclopaedic attitude capable of keeping together heterogeneous elements appears to me to be requested by the object itself of semiotics, sense. If semiotics really wants to be the discipline of systems and processes of signification and communication it must indeed deal with a double burden, which is instead often escaped: (i) sense is an irreducibly complex phenomenon, i.e. dependent on a multiplicity n of parameters that often contradict each other[16]; (ii) sense is not something that circulates exclusively within the domain of semiotics, which can thus ascribe it as its own object, rather it is something that circulates within all disciplinary domains that belong to a semiosphere, since every disciplinary domain is built on autonomous types of investigation, appropriateness, sense and truth that establish their own systems of signification and communication.[17]

Point (i) thus leads to the construction of a semiotics of complexity, irreducible to those features of coherence, closure and cohesion that were thought to constitute the semiotics of text and Hjelmslev's idea of structure. Instead, point (ii) leads to the construction of an encyclopaedic semiotics, able to recognise that sense is the condition of possibility of the existence of any disciplinary domain, and not the exclusive object of semiotics. Hence an essential corollary concerning point (ii): if semiotics wants to properly account for the systems and processes of signification and communication, it must always live in translation with an «outside», *en plein air*, moving between the practices of signification within the encyclopaedia. Hence its being an interpretive and translational logic. Hence the existence of semiotics of arts distinct from artistic theories, of semiotics of perception distinct from the study of perception, of semiotics of music distinct from the study of music, of semiotics of literature distinct from literary analysis, etc. Semiotics lives on translation with an «outside», because its «inside» is the encyclopaedia (cf. Eco 1984) and the

---

[15]Cf. La Mantia (2012).

[16]This is Morin's definition of complexity (1986: 36–40).

[17]Cf. Serres (1968, 1972).

semiosphere (cf. Lotman 1985), which by definition are systems composed of disciplinary systems that are different and heterogeneous to one another.

Hence our most complete thesis:

1. Sense constitutively depends on both immanent dimensions (semio-linguistic structures) and dimensions that transcend semiotics (perception, cognition, culture, etc.).
2. All transcendent dimensions are also complex structures that depend on multiple parameters, often in contradiction to one another (cf. Morin 1986: 36–40).
3. If semiotics does not wish to confine itself to pure immanent discipline, dealing only with the second dimension of Saussurean value (and thus leave itself open to the criticisms directed towards it), it must integrate variables with point 1 and think of them as constitutive of its object (sense). This was for instance Charles Sanders Peirce's lesson, who established semiotics precisely from a reflection on cognition connected to a theory of perception and interpretive habits within the interpretant community.

In doing so, descriptive structures of semiotics must be thought as constitutively complex, i.e. able to describe phenomena that are largely dependent on potentially contradictory variables that are constantly evolving, such as cultures, social systems or cognition.

For this reason, I believe that encyclopaedic semiotics is a complexity semiotics, able to keep together heterogeneous systems, recognising the existence of several good reasons (cf. Morin 1986).

5.6 Conclusion

Recently, Paolo Leonardi highlighted how between 1975 and 1990 the «linguistic turn» that marked analytical as well as continental philosophy (from Wittgenstein to Austin to Searle on the one hand, from Heidegger to Ricoeur to Gadamer on the other), was overcome thanks to the adoption of Chomsky's generative linguistics as a model for cognitive sciences. But since it is precisely cognitive sciences that reject Chomsky's model and open themselves to complexity theory, it should be remembered how between 1975 and 1990 it was precisely the complexity theory to revive those same systemic principles that were at the basis of the semio-linguistic turn as I have described it in this work. For my part, semiotics will become central to contemporary cultural discourse when it will be able to present itself as a discipline of complexity, able to offer a theory of cognition that goes beyond the dichotomies between nature and culture, object and concept, fact and representation, moving on that thin line between *physis* and *nomos* that, as Morin

(1977, 1993) taught us, is «more real as it is more rooted in physis and connected to it» than «the old, almost artificial object and its pseudo-realism».

Beyond any welcome back we can say to reality, beyond any project of naturalisation founded on reductionism as well as beyond any «logic of cultures» founded on a purely descriptive apparatus, there is the bond between complexity theory, semio-linguistic systems and theory of cognition that contemporary semiotics must learn how to express in its own way.

References

Carta, G. (2012). *Rizoma e sistemi complessi*, Tesi di laurea in Semiotica interpretativa, Università di Bologna.

Coquet, J. C. (2007). *Physis et Logos. Une phénoménologie du langage*. Paris: Presses Universitaires de Vincennes.

Deleuze, G., & Guattari, F. (1980). *Mille plateaux. Capitalisme et schizophrénie*. Paris: Les Éditions de Minuit.

De Saussure, F. (1967). *CLG Cours de linguistique générale*. Lausanne-Paris: Payot.

De Saussure, F. (2003). *ELG Écrits de linguistique générale*. Paris: Gallimard.

Eco, U. (1968). *La struttura assente*. Milano: Bompiani.

Eco, U. (1975). *Trattato di semiotica generale*. Milano: Bompiani.

Eco, U. (1978). Il pensiero semiotico di Jakobson. In: R, Jakobson (Ed.), *Lo sviluppo della semiotica* (pp. 5–32). Milano: Bompiani.

Eco, U. (1984). *Semiotica e filosofia del linguaggio*. Torino: Einaudi.

Eco, U. (2007). *Dall'albero al labirinto*. Milano: Bompiani.

Fabbri, P. (2005). *La svolta semiotica*. Roma: Laterza.

Ferraris, M. (2012). *Manifesto del nuovo realismo*. Roma-Bari: Laterza.

Ferraris, M., & e De Caro, M. (2012). *Bentornata realtà*. Torino: Einaudi.

Fontanille, J., & Zilberberg, C. (1998). *Tension et signification*. Liège: Mardaga.

Gandolfi, A. (2008). *Formicai, imperi, cervelli. Introduzione alla scienza della complessità*. Torino: Bollati Boringhieri.

Greimas, A. J. (1970). *Du sens*. Paris: Seuil.

Greimas, A. J. (1983). *Du sens II. Essais sémiotiques*. Paris: Seuil.

Greimas A. J., & e Courtés J. (1979). *Sémiotique. Dictionnaire raisonné de la théorie du langage*. Paris: Hachette.

Hjelmslev, L. (1959). *EL Essais linguistiques, Travaux du Cercle Linguistique de Copenhague*, vol. XII.

Hjelmslev, L. (1975). *R Résumé of a Theory of Language, Travaux du Cercle Linguistique de Copenhague*, XVI, edited and translated with an introduction by J. Francis. Whitfield: The University of Wisconsin Press.

Hjelmslev, L. (1985). *NE Nouveaux essais*. Paris: PUF.

La Mantia, F. (2012). *Che senso ha? Polisemia e attività di linguaggio*. Roma: Mimesis.

Lotman, J. (1985). *La semiosfera*. Venezia: Marsilio.

Licata, I. (2008). *La logica aperta della mente*. Torino: Codice.

Licata, I. (2012). Epistemologia adattativa: vedere con i modelli. *Reti, Saperi, Linguaggi, IV, 1*, 9–16.

Morin, E. (1977). *La Méthode 3. La nature de la nature*. Paris: Seuil.

Morin, E. (1986). *La Méthode 3. La connaissance de la connaissance*. Paris: Seuil.

Morin, E. (1993). *Introduzione al pensiero complesso*. Milano: Sperling and Kupfer.

Paolucci, C. (2010). *Strutturalismo e interpretazione*. Milano: Bompiani.

Paolucci, C. (2012). Physis e nomos. Ideologie della natura tra catarsi, empatia e percezione sessuale. In G. Marrone (Ed.), *Semiotica della natura* (pp. 79–102). Milano: Mimesis.

Peirce, C. S. (1958). **CP** *Collected Papers of Charles Sanders Peirce.* voll. I–VI edited by C. Hartshorne & P. Weiss (1931–1935), voll. VII–VIII edited by A. W. Burks (1958). Cambridge, (Mass): Belknap Press.

Peirce, C. S. (2009). ***W*** *Writings of Charles Sanders Peirce. A Chronological edition.* Bloomington: Indiana University Press.

Petitot, J. (1979). *Locale/globale.* Torino: Enciclopedia Einaudi.

Petitot, J. (1985). *Morphogenèse du sens* (Vol. 1). Paris: PUF.

Rastier, F. (2004). Deniers et veau d'or: des fétiches à l'idole, http://www.revue-texto.net/Inedits/Inedits.html.

Serres, M. (1968). *Hermès I. La communication.* Paris: Les Éditions de Minuit.

Serres, M. (1972). *Hermès II. L'interférence.* Paris: Les Éditions de Minuit.

Zilberberg, C. (1983). *Raison et poétique du sens.* Paris: PUF.

Zinna, A. (2001). *"Il concetto di forma in Hjelmslev", in Janus (2).* Padova: Imprimitur.

Chapter 6
Hjelmslev and the Stratification of Signs and Language. The Morphodynamical Approaches to Semiotic Complexity

Antonino Bondì

6.1 The Stratification of Language Between Sedimentation and Metamorphosis

According to Hjelmslev (1954), the theory of linguistic and semiotic stratification is a way that aims at the convergences of the connections among linguistic activity, natural languages and speech acts into a dynamical system. Hjelmslev think that the relationship between the internal dynamics of the system components and the variations of semiotic forms, set by the social change and its heterogeneity, should be considered in the terms of an osmosis, by refocusing the virtuous circle between use and collective transmission of linguistic and semiotic forms. Furthermore, Hjelmslev's theory of linguistic object is based on this foundational circularity: the languages seem to constitute internally dynamic concrete systems, regulated, and organized on the basis of constraints and dependencies. Constraints and dependencies are the variety of linguistic objects. Such variety reveals itself through a multiplicity of local morphologies, driven by two correlated metamorphic pressures: (1) internal pressures, related to system of dependencies and (2) external pressures, folded up to the constraints that speaking communities impose.

Thus, it is better to identify the layers of linguistic forms by the means of different stages of the «life» of linguistic objects. From this point of view, Rastier (2006) has argued that Hjelmslev project was based on decoupling semantics from a naively realist metaphysics and on emphasizing the transformation and the local changes in language in a panchronic framework, i.e., the study of language through natural languages. According to Rastier, this is where:

> une théorie des formes linguistiques débouche nécessairement sur une théorie de transformations, car une forme n'est qu'un moment de stabilisation dans une suite de

A. Bondì (✉)
LIAS, EHESS, Paris, France
e-mail: antoninobondi80@gmail.com

F. La Mantia et al. (eds.), *Language in Complexity*,
Lecture Notes in Morphogenesis, DOI 10.1007/978-3-319-29483-4_6

> transformations. Cela vaut pour les transformations au sein d'une performance linguistique, par exemple les transformations thématiques entre le début et la fin d'un texte, mais aussi entre performances, entre textes séparés dans des séries temporelles. (Rastier 2006: 96).

From this point of view, Hjelmslev's theory is a theory of the semio-linguistic event: an intersection of strata or layers that social and linguistic mind sediments and make latent. The constitutive instability of system makes strata constantly reusables:

> On peut définir la parole par la rencontre même et l'entrecroisement des strata. La parole en effet est, en dernière analyse, tout ce qui est arbitraire dans le langage. La parole se définit comme l'ensemble des relations interstratiques effectivement exécutées. L'usage, à son tour, est évidemment ce qu'il y a de stabilisé dans la parole. L'usage se définit comme l'ensemble des connexions interstratiques effectivement exécutées (Hjelmslev 1954: 76).

In order to define the speech activity (parole) as an intersection or a weaving of strata, it is necessary to rethink the relationship between langue and parole as dynamical relations between inseparable modalities of language. These modalities are bound to the pressure of time and the variations that time requires, as well as to the typologies of forms and operations taking part in its formation and deformation. In this context, Hjelmslev theoretical contribution to a theory of the enunciative praxis can really be appreciated.

With regard to the semantic aspect of the theory, language, and semiotic stratification makes us face up to a multiplicity of overlapping instances of categorization. Hence, sign is a dual unity that arises from emergence of a dynamism oriented by mutual control relations between phonetic and semantic spaces. Moreover, from a phenomenological point of view, the semiotic substance is dense because it results and appears from recognition layers brought into play by social minds. Among these layers of recognition, the collective appreciation coordinates and animates cultural practices, semantic forms, axiological systems, values, etc.: «the sociobiological conditions and psycho-physiological mechanisms […] allowing to the speaking subjects […] to create, reproduce, evoke, and handle in different ways the elements of appreciation» (Hjelmslev 1954: 61–62). From this point of view, describing language in terms of a complex system seems a fruitful way. As «complex system» we mean a totality of elements and processes interconnected by a network of relationships, and whose organization is structured by layers. These layers produce a series of interactions, through which, for each level, almost constantly original properties can emerge, and which are not explainable starting from the lower levels. On the contrary, the set of relationships between global system and local processes is the only one that allows to describe the organization and the dynamics of the system: in a word, its semiotic life.

6.2 A Sign Beyond the Sign?

Hjelmslev has always criticized the saussurean postulate of indissolubility of signifier and signified, and has established the need to start from the text—considered as a genuine phenomenological data—and to subdivide it analytically. A set of figures of expression and figures of content derives from the methodological text breakdown going from semiotic performances to minimum sizes of signs. Therefore, the linguistic analysis should not aim at the study of signs—and of language as a system of signs—but at the analysis of phonic and semantic elements, which not necessarily coincide with the sign. If languages are multifunctional patterns of relationships between elements, a notion of sign linked to a lexicalist vision is inadequate. The «dependencies» as semantic and phonic unities identified by the analysis are not structured by one-to-one correspondences. For justifying his descriptive practice, Hjelmslev needs to break up the connection between signifier and signified, that is the ultimate limit of analysis. According to Ferdinand de Saussure, the language is «un système de signes où il n'y a d'essentiel que l'union du sens et de l'image acoustique» (de Saussure 1922: 63); the union of sound and meaning (or sense) constitutes the mean characteristic that determinates the objectivity of linguistic forms: «nous appelons signe la combinaison du concept et de l'image acoustique (…). Nous proposons de conserver le mot signe pour désigner le total, et de remplacer concept et image acoustique respectivement par signifié et signifiant» (de Saussure 1922: 99). The signifier, as well as the significant, and their organized totality—as the three constitutive dimensions of sign—are grounded on differential relations, mutual references, and oppositions. As David Piotrowski has noticed, neither for Saussure nor for Hjelmslev it is sufficient to stop at this stage of analysis:

> pour atteindre une entière compréhension structurale il reste encore à renverser l'ordre des primautés, c'est-à-dire à établir l'existence du signifiant et du signifié sur le principe même de la liaison organique qu'ils contractent et qui les institue dans leur qualité de grandeur linguistique. À ce moment, signifiant et signifié ne sont plus seulement les appellations, mieux appropriées, des objets acoustique et sémantique insérés dans l'ordre linguistique par l'effet d'un rapport indéfectible et totalisant, mais des fractions de contenu et d'expression construites sur le principe d'une fusion (Piotrowski 1997: 118–119).

According to Saussure, the linguistic entity is the association between a signifier and a significant; when one considers only one of two categories, the linguistic unity disappears. The morphogenetic dimension of language intertwines to the psychological dimension but remains bound to a new system of configuration of forms. This system can not be reduced to prelinguistic dimensions. Any form of division between signifier and significant means to drop out the linguistic reality. On the contrary, Hjelmslev intends to overcome the union of sounds and senses; it needs to create two planes in which the series of phonics and semantics quantities are variously organized and arranged, in relation to each other. For this reason, Hjelmslev develops in an original way the notion of form. Like Saussure, Hjelmslev believes in the primordial nature of the semiotic bond between acoustic and semantic regions. However, he reorganizes the model of signs and suggests the analysis of connections of expression and content planes. The stratification theory results from a methodological and analytical passage, i.e., from sign to semiotic function. Functional connection is an operational system, whose intersection generates inherently stratified objects. Despite differences, this theoretical view is deeply saussurean:The sign function is in itself a solidarity. Expression and content are

> solidary-they necessarily presuppose each other. An expression is expression only by virtue of being an expression of a content, and a content is content only by virtue of being a content of expression. Therefore—except by an artificial isolation—there can be no content without expression, or expressionless content; neither can there be an expression without a content, or content-less expression. If we think without speaking, the thought is not a linguistic content and not a functive for a sign function. If we speak without thinking, and in the form of series of sounds to which no content can be attached by any listener, such speech is abracadabra, not a linguistic expression and not a functive for a sign function (Hjelmslev 1961: 48–49).

The transfer operated by Hjelmslev from sign to semiotic function shows an original reorganization; indeed, semiotic and linguistic stratification theory aims to describe connections between morphogenesis of forms and their underlying operations. At the conclusion of La stratification du Langage (1954), Hjelmslev clarifies these links between variability of social use implied by linguistic unities and operational layers realized by the speaking communities:

> Nous croyons en effet que les réflexions que nous avons maintenant terminées permettent de donner une définition non seulement de l'usage, mais aussi, plus généralement, de la parole, et, d'autre part, plus particulièrement, de la norme (…). On peut définir la parole par la rencontre même et l'entrecroisement des strata. La parole est en effet, en dernière analyse, tout ce qui est arbitrare dans le langage. La parole se définit comme l'ensemble des relations interstratiques effectivement exécutées. L'usage, à son tour, est évidemment ce qu'il y a de stabilisé dans la parole. L'usage se définit comme l'ensemble des connexions interstratiques effectivement exécutées (…). À la différence de l'usage, la norme doit être l'ensemble des relations interstratiques admises. Le schéma sémiotique (et linguistique) par contre est en dehors de cet ordre d'idées: ce terme ne se rapporte qu'aux fonctions intrinsèques dans la forme de chacun des plans pris à part (Hjelmslev 1954: 76).

Inside this logic, the fundamental principles of hjelmslevian linguistic can be understood: they aim to create a retrieval strategy of continuous and immanent transformations of language activity through different languages.

6.3 The Sign as Accident: Toward the Semiotic Function

Shifting the focus to the concept of semiotic function, Hjelmslev means to describe the scheme of sense organization and intends to fight against the metaphysics of componential models of meaning. Componential theories conceive languages in the terms of closed, abstract, and static systems; these models can not describe lexical meaning, as well as the experience of sense, such a «head of Meduse», according to Benveniste. As Jacques Fontanille noticed (Fontanille 2004: 19–28), the evolution of the notion of semiotic function shows that the position of the danish linguist is certainly less clear than anyone expected. In a first moment, the semiotic function has been understood as a logical relationship of presupposition between elements of expression and content planes, without any reference to the operators of this connection—the body and the enunciation—nor putting in account those factors that make the dynamic function of reciprocity as essential as fragile and mobile.

However, if we try to account for the complex and stratified dynamics that organize the linguistic structures, this interpretation of semiotic function is a non sufficient; such an interpretation has removed the dimensions which actually constitute the focus of linguistic and cognitive debate: body; cognition; intersubjectivity of sense; temporality; enunciation. This theoretical removal has been legitimated by the notion of sign that is the heart of structuralist and saussureian linguistic tradition. Following, a logic-formalist and disembodied conception of sign, this tradition cannot explain the complexity of «speaking bodies» and «enunciative structures», as well as the linguistic mind. However, the hjelmslevian theory of semiotic function enables a reflection both on the status of the bond between expression and content and on the status of the semiotic-cognitive operations that generate this duality. In this way, it is possible to go back to regimes of duality or oneness of linguistic objects.

This connection animates the reinterpretation of hjelmslevian theory as well as the status of semiotic and linguistic objects and finally the languages as formal and relational systems. Languages are shaped by linguistic objects and their dynamical relations, as well as by different semiotic and cognitive operations. These three layers—objects, languages, and semiotic operations—overlap and reshape constantly; the concept of the stratification intends to give an account of their relations. Thus, Hjelmslev's stratification theory is a general semiotic hypothesis that is able to grasp the tangle of operations and layers of semiotic substances determining such a multiplicity of layers and multiple concrete performances describing the ecology of human kind. We suggest an ecological interpretation of Hjelmslev's semiotic theory of stratification. We aim to establish an ontology of stratification of linguistic objects and their constitution processes, i.e., the langue and the parole as two interactive modalities from which the activity of language can emerge. Thus, the language praxis can be represented as a work of creation, stabilization and deformation of social and individuals forms in the collective networks of enunciation. We will present such a topological and dynamic semiotics, which identifies the ecological stratification of sense. At the core of the hjelmslevian semiotic topology, we find the collective and individual «usages» that constitutes the keystone of theory of the semiolinguistic events.

6.4 From Translatability of Sense to Opacity of Languages

Both in the Prolegomena to a Theory of Language (1943) and the Stratification of Language (1954), Hjelmslev defines the linguistic system as «omniformative». In particular, the question is discussed in relation to translatable nature of semiotic and especially linguistic systems:

> A language may be defined as a paradigmatic whose paradigms are manifested by all purports, and a text, correspondingly, as a syntagmatic whose chains, if expanded indefinitely, are manifested by all purports. By a purport we understand a class of variables which manifest more than one chain under more than one syntagmatic, and/or more than one

> paradigm under more than one paradigmatic. In practice, a language is a semiotic into which all other semiotic may be translated—both all other languages, and all other conceivable semiotic structures. This translatability rests on the fact that languages, and they alone, are in a position to form any purport whatsoever; in a language, and only in a langue, we can work over the inexpressible until it is expressed (Hjelmslev 1968: 109).

A deep conviction is rooted in saussurean tradition, according to which the meaning is constitutively bound to the signifier through the function of the sign, so that every change—deformation or substitution—concerning the significant, regard at the same time the inclusive meaning. Therefore, a change of significant can produce a transformation or a disappearance of whole sign. Hjelmslev partially shares this view, and he emphasizes that saussurean conception of sign conceived as an organized totality contrasts with the traditional interpretation of sign, according to which the sign is a sign of an extrinsic entity, that is the referent:Up to this point, we have intentionally adhered to the old tradition according to which a sign is first and foremost a sign for something. In this, we are certainly in agreement with the popular conception and, moreover, with a conception widely held by epistemologists and logicians. But it remains for us to show that their conception is linguistically untenable, and here, we are in agreement with recent linguistic thinking. While, according to the first view, the sign is an expression that points to a content outside the sign itself, according to the second view (which is put forth in particular by Saussure and, following him, by Weisgerber) the sign is an entity generated by the connexion between an expression and a content (Hjelmslev 1968: 47).

Apparently, following the suggestions from the 13th paragraph of Prolegomena, it seems that Hjelmslev proceeds in a completely opposite direction:There will always be solidarity between a function and (the class of) its functives: a function is inconceivable without its terminals, and the terminals are only end points for the function and are thus inconceivable without it. If one and the same entity contracts different functions in turn, and thus might apparently be said to be selected by them, it is a matter, in each case, not of one and the same functive, but of different functives, different objects, depending on the point of the view that is assumed, i.e., depending on the function from which the view is taken. This does not prevent us from speaking of the «same» entity from other points of view, for example, from a consideration of the functions of that enter into it (are contracted by its components) and establish it. If several sets of functives contract one and the same function, this means that there is solidarity between the function and the whole class of these functives, and that consequently each individual functive selects the function. Thus, there is also solidarity between the sign function and its two functives, expression and content. There will never be a sign function without the simultaneous presence of both these functives; and an expression and its content, or a content and its expression, will never appear together without the sign function's also being present between them (Hjelmslev 1968: 48).

From this point of view, the sign is a rigid entity; every sense transformation implies a total change of sign's structure. Despite the appearances, Hjelmslev contrasts with this perspective. Theorizing the form of content as autonomous and parallel to the form of expression, the Danish linguist suggests an interpretation of saussurean sign allowing to think the relationships between expression (significant) and content (meaning) in the terms of a relation between two forms, that are morphologies. According to Per Aage Brandt, the hjelmslevian model recognize «l'existence d'une plasticité sémiotique à l'endroit du signe, puisque ses deux formes émergent de deux substances et ne peuvent pas se constituer mutuellement (…). Et, si chaque forme gère sa propre stabilité, le signe se présente comme une bi-stabilité dont l'identité sera compatible avec des déstabilisations relatives, aux deux niveaux» (Brandt 1994: 83). This semiotic plasticity can be analysed through the deformations and parallel transformations of both semiotic planes, because of dynamic and bi-stable nature of sign. Following Brandt:

> La question de l'identité du signe est ainsi à reprendre dans la perspective du trasfert de sens; et en premier lieu, la question de l'identité du mot est à repenser; l'échange et l'apprentissage, la transmission en général des signes établissent une telle identité à travers une double interprétation, celle de la forme de l'expression et celle, autonome, de la forme du contenu. Pour qu'un signe – tel un mot – soit identifié, il faut précisément que ce qu'il signifie puisse être signifié autrement (…). Le signifiant est ce qui renvoie son signifié à un autre signifiant (…). Le même se constitue à travers l'autre. Ainsi, la traduction entre dans la définition même de la signification (Brandt 1994: 83).

From this point of view, the translatability of sense tests the «monological» principle behind Saussure's intuition. When a transfer of sense occurs, a form of content disengages from a original form of expression to drift into a new form. Thus, there are two ways of considering this phenomenon. The first way—less relevant—is represented by a structuralist representation of sign. It establishes a kind of relationship between two substances of the content with respect to the identity of matter that is anterior and external to every linguistic problem. Sense transfer underlying the interlinguistic translatability is considered, in this perspective, as a secondary effect. But there is a second more interesting way to approach this problem. As written by Brandt, Hjelmslev starts from the translatability of facts, experiences and forms—conceived as constitutively uncertain—and searches for linguistic invariants in order to explain the continuous variation of sense. In this way, we highlight the presence of a line of research aiming to rewrite the notions of semiotic function and semiosis; this second theoretical tendency wants to be able to account the intrinsic semiotic translatability that Hjelmslev sets at the basis of definition of language. If we are deprived of link between language, translatability, and stratification, it becomes difficult to understand the anthropological idea of language that Hjelmslev's structuralism recommends. Thus, this second way can undertake two branches. The former is the assumption of the centrality of the substance, and the consequently substantialist approach to the emergence of meaning phenomenon. The latter is the maintenance of the hypothesis about two constitutive planes determining the building of semiotic forms.

Therefore, as Brandt observes, it is necessary to assume, in order to defend the two planes of the sign's hypothesis, that what exists are «des morphogènes, qui font émerger les formes dans la substance spécifique de chaque plan» (Brandt 1994: 84).

Only in this way, the difference of translatability of semiotic forms is not reduced to a quantitative fact but, on the contrary, it is inherently qualitative, linked to the stratification of the categories that come into play in the social life of forms, and bound to the relation of opacity between languages and meaning experience.

6.5 The Morphologies of Sign and the Problem of Traduction

The exclusion of the substance has been one of the main problems of Hjelmslev's theory. This exclusion, when we consider an essay as The Stratification du Langage, is unclear. Hjelmslev places the last tasks of linguistic theory beside the analytic-deductive procedure:

> Vouloir comprendre dans la forme linguistique celle du contenu, et non seulement celle de l'expression; et, en conséquence de ces principes, 4° celui de considérer le langage, dans le sens communément adopté par les linguistes, comme un cas particulier d'un système sémiotique, c'est-à-dire d'un système comportant des plans différents et, à l'intérieur de chaque plan, une différence entre forme et substance (…) et de situer la linguistique dans les cadres d'une sémiotique (ou sémiologie) générale (Hjelmslev 1954: 46–47).

In order to understand semantic phenomena and their natural transmissibility we have to configure the centrality of substantial dimension and the relationship of mutual control and coexistence between the two planes. For this reason, the theory of stratification tries to keep together the layers or strata immanents to the construction of the sense, and the substantial layers: referenciation, semantic domains (historical, cultural, anthropological, cognitive, etc.). Here, we are talking about usage, that is the foundation of the stratification of content. The restriction of languages to their formal dimension (with the consequent exclusion of the substance) represents only a step in the analytical procedure, related to the holistic nature of linguistic phenomena. On the other side, Hjelmslev writes that one of the fundamental requirements for the comprehension of sense phenomena is the focus on the multiplicity of substances. He argues that the same theory of stratification would make no sense if one does not think this multiplicity as the basis of integral nature of the substance of the content. This is the precondition for translation practice and, more deeply, for translatability as an intrinsic property of the linguistic meaning:

> En combinant le vues traditionnelles sur la délimitation de la linguistique avec l'analyse fonctionnelle entreprise par la glossématique, tout le monde sera d'accord pour reconnaître que le domaine dévolu à la linguistique, le domaine qui lui est réservé et qui constitue son objet propre et spécifique, est constitué par l'ensemble des unités dont la plus large est la phrase et dont la plus petite est le glossème. Il est vrai que cette ligne de démarcation ne doit avoir rien d'absolu, et que par contre le point de vue élargi introduit par la glossématique et la nécessité d'une description exhaustive exigent non seulement que dans la description la hiérarchie de la forme, épuisée par l'établissement des glossèmes, soit accompagnée, s'il y a lieu, des hiérarcies des substances, mais aussi que l'analyse circonscrite par la ligne de démarcation indiqueée soit précédée d'une analyse dégageant les unités plus larges (…). C'est ainsi que la sémiotique réclame le droit d'une discipline qui puisse servir de norme pour toutes les sciences de l'homme (Hjelmslev 1954: 66–67).

In this context, translation/translatability becomes the keystone of this theoretical structure. The translatability determines the morphologies of semiotic system and the relations between morphologies and substantial domains, i.e., the concrete realizations in discursive and enunciative experience. The centrality of translatability appears in a paper dedicated to the study of semantics:

> Des langues européennes qui nous sont familières distinguent entre frère et soeur; le malais cependant ne fait pas cette distinction et possède un seul signe ou mot signifiant indifféremment «frère» et «soeur», si bien que seul le contexte ou la paraphrase peuvent indiquer avec certitude laquelle des deux variants sémantiques est manifestée. D'autres langues ne se contentent pas de distinguer entre frères et soeurs, mais distinguent aussi cadets et aînés. Ainsi le chinois et le hongrois, par exemple, ont quatre mots correspondant à nos deux mots frère et soeur (…). La forme du contenu de ces langues établit plus de distinctions à l'intérieur de la substance du contenu que ne le font nos langues. Nous découvrons aussi qu'une seule et même zone de substance dans le contenu est mise en forme de différentes manières dans des langues différentes, si bien que, quoiqu'il y ait des distinctions dans deux langues, elles sont établies à des endroits différents et ells entretiennent donc des relations décalées. Un example bien connu en est la relation des deux mots français arbre et bois avec les deux mots danois troe (arbre, bois) et skov (bois, forêt). La ligne qui sépare arbre et bois n'est pas la meme que la ligne qui sépare troe et skov: la variante sémantique «bois en tant que matériau» par opposition à «bois en tant que plante» prend en français la forme d'une variante contenue dans le mot bois, et en danois celle d'une variante dans le mot troe (Hjelmslev 1971: 101).

Remarks about differences and demarcation lines realized by the different languages are one of main theoretical fields where problems of traduction play. Moreover, «c'est seulement grâce à la théorie proposée par Saussure et ses disciples que cette observation a été correctement interprétée. Leur point de vue a certes de consequences capitales pour la sémantique. La forme sémantique n'est pas distincte de la langue; au contraire, elle est une part importante de la langue elle-même» (Hjelmslev 1971: 101).

According to Hjelmslev, there is no semantics outside the linguistic domain. Variations of demarcation that languages require have to be observed by following the emergence of processes from some substances of content (concrete manifestations of intersubjectively adjusted and shared semantic uses) that are pre-structured, organized, and already immersed in the field of the anthropological history and community semantics. For this reason, the translation takes part in definition of meaning:

> On s'est depuis longtemps aperçu que, si profond soit le fosse qui sépare les langues, elles peuvent en venir à se ressembler s'il existe entre elles une communication culturelle (…). Antoine Meillet a également trouvé que les langues ayant une origine commune, bien qu'elles se développent dans des directions divergentes, peuvent, au cours de ce développement, montrent certains traits parallèles, comme conséquence de leur environnement culturel commun. Et il a montré que les langues européennes, particulièrement celles de l'Europe et de l'Ouest, partagent un grand nombre de traits que l'on peut difficilement attribuer à leur origine commune, ou à l'influence culturelle traditionnelle du grec ou du latin, mais qui doivent dériver d'un mode de vie homogène (…). Les révolutions linguistiques récentes ont produit des associations linguistiques d'une très large étendue géographique, avec des systèmes de signes politiques communs, une terminologie politique commune, ou «idéologie» (…). Le fameux «Rideau de fer» est une frontière sémantique entre deux énormes associations linguistiques, chacune avec sa forme spécifique du contenu dans la totalité de cette zone de substance qui peut être appelée le politique au sens plus large. Le défaut de compréhension entre ces deux mondes est, en dernière analyse, et sit ant qu'on puisse le surmonter, une question de traduction (Hjelmslev 1971: 103).

The semiotic idea of translation shows some concepts, such as the homogeneous way of life, cultural environment, semantic border, and substance region. Why?

Translation—or translatability—actuates the dynamics defining sense formality, its shaping forms provided with an internal plasticity activity; formal demarcations of languages should be conceived as emergence of internal conflicts substances.

6.6 The Three Morphologies: Grammatical, Expressive, Semantic

Now, if we go back to the problem of translatability as main property of semiotic forms, the theme of an accurate description of what we mean by morphologies and by their tasks become immediately necessary. We analyze the status of the word from the point of view of the types of variation that it achieves. Their inquiry unfolds analytical operations allowing an approach to their functioning. First of all, let's get a first morphology that is not a real investigation subject for us, but that we recapitulate, according to Per Aage Brandt. In consonance with Hjelmslev's conviction—rooted and disseminated in his writings—the distinction between lexical and grammatical meanings would prove an impossible task. Brandt draws a diagram of co-variation relationships of expression and content, with a view to problems of translation as a natural trait of morphogenesis, movement, and transmissibility of the sign. Whereas, the morphologies of the expression and the content exhibit trend and preference lines, which govern and constrain the semiotic change, as well as their invariance, this kind of morphologies stabilizes languages schematically and cognitively. Morphologies constitute a set of operators and domains of meaning «qui se prêtent à cette mise en forme particulièrement forte et prégnante» (Brandt 1994: 88), i.e., the morphological structure of languages, with their interlinguistic shifts. Supposing, for example, that a continuous variation at meaning level correspond to a continuous variation equally situated at significant level. In hjelmselvian terms, there is parallel change between the two planes. In this case, according to Brandt, the switchings on meaning and significant levels lead to the same result. Let us take, for example, the french word cheval, and let us just consider the variation from cheval to chevaux. It is intuitive to observe that this variation on the significant plane causes a variation on the content plane, i.e., variation between the singular and the plural. In this case, we are dealing with grammatical morphologies, whose the properties are: (1) a regime of co-variation of two planes and of iconic representation between significant and signifier; (2) a small amount of these formations against a higher frequency in language expressions; (3) the schematic nature of morphological formations. These domains pervade the entire linguistic system; they are cognitive structures and universal as patterns of content: time, modality, causality, aspect, enunciative position, etc. In conclusion, this first kind of morphological structure in languages is the first thermometer of a cognitive and interlinguistic work, that speaking subjects realize in constructing the linguistic system. The structure of morphological languages structure cannot be conceived as a formalist framework of relationships between discrete elements, but is a morphological domain that regulates a system of phonic/semantic and syntactic variations. That makes the notion of structure more dynamic and plastic:

> les catégories morphologiques (...) sont générales, elles ne sont pas universelles. Elles ne sont pas réalisées dans le système de n'importe quel état de langue, mais elles résident dans le système du langage à titre de possibilités. Ces catégories sont mutuellement autonomes, non complémentaires, c'est-à-dire même si les catégories ne se réalisent pas toutes dans une langue donnée, on peut déterminer celles des catégories qui s'y réalisent; même si au cas extreme, une seule catégorie se réalise, on peut en constater l'existence et identifier la catégorie réalise par rapport au système du langage (...). Les catégories ou corrélation dont nous avons parlé jusqu'ici sont donc les catégories ou corrélations préétablies du langage, préexistant à la langue, générale et réalisables. Ce sont les combinaisons et les déterminations qui les réalisent dans une langue donne (...). Ces catégories et corrélations sont d'une nature différente. Sans préexister dans le langage, elles s'établissent dans chaque langue prise à part. Ce sont les défectivations et les simplifications qui les établissent dans une langue donnée (Hjelmslev 1971: 172–173).

We consider now the hjelmslevian concept of commutation, which is the analytic operation identifying relevance criterion in order to define variation in the study of expression. We follows once again the example of Brandt, i.e. the french word cheval. We suppose to understand with this 'X' the animal equus drawn by Ferdinand de Saussure in the Cours de Linguistique Générale. Now, we suppose to apply a deformation in pronunciation to word, getting chevaul. As Brandt emphasizes, comprehension can rest the same, even with various degrees of uncertainty or lower stability; these are cases in which strangenesses in flexion, in intonation or phonetic accidents may occur, but they are tolerated by linguistic community perception. But, let us just imagine that the deformation continues and produces a word like chevoul, too remote from starting word. At this point, the meaning disappears, because of an exasperation of tolerance in perceptive community. In this case, the pronunciation exceeds a threshold; the gap between a-ou is finally distinctive. Hjelmslev has defined the commutation as a method to identify distinctive forms by studying the effects on the two planes through local and substantial variations:

> La variation locale, qu'elle soit productrice ou destructrice, nous permet ainsi, au niveau de l'expression, de découvrir, dans la substance, des écarts distinctifs établissant une forme. On pourrait proposer de préciser en parlant, en général, ici, de commutations vers le signifié. Elles nous permettent d'établir, dans une langue, des morphologies expressives (signifiantes). D'ailleurs, les variations qui rencontrent et traversent des frontières distinctives aboutissent bien plus souvent à des disparitions du signifié qu'à l'apparition d'un nouveau sens, le signifiant du mot – et probablement de tout signe – semble ainsi entouré d'un certain vide (...), comme si le signe était protégé, dans son identité, par un environnement combinatoire de non-mots, par une auréole de variantes vides (Brandt 1994: 86).

This is a delicate point for the theory of semiotic stratification. In order to understand what we mean by «expressive morphologies», it is necessary to relate the problem of variations to Hjelmslev's theoretical triad: form, substance and matter. These are correlatives termes: there are no forms, no substances, no matters, no contents or expressions in themselves, but they coexist in classes of mutual relations. Thus, the study of variation allows to approach differently the theory of stratification and to understand the topological and dynamical intuition of bio-cognitive structuralism. With reference to the study of expression, these examples of continuous variation suggest that the whole «singular points» of internal states define the form of expression, by constituting morphologies of expression. These are local systems, like phonic Gestalten, that allow a continuous variation within a heterogeneous space, without losing sight of invariants of Gestalt's organization itself, i.e., its points of catastrophe. On one hand, morphological space of expression is stratified according to the set of relationships that phonemes undertake, and, on the other hand, according to the emergencies of perceptive discontinuities in substances of the expression; this space is distributed on the basis of the coexistence of forms, substances and matters of the expression. That allows us to try to form a first sketch of the structure of expression. The form of expression, far from being an abstract set of phonemes, has to be identified as an emergency on the material plane of a tangle of local expressive morphologies that govern the expressive variations of meaning. These morphologies, when they emerge as formal discontinuity, are immediately projected into an external control space (the substantial domains), where phenomenal world appears.

Following the suggestion of Brandt, let us formulate the inverse operation, rebuilding the paradigmatic stratification of the content. In other words, we start now from the meaning of our word cheval, and observe its semantic variations within the quadrupeds field. So, we get terms as centaure, hippogrife, licorne, pégase, étalon, jument, poulain, etc. As pointed by Brandt, we are in the presence of a variation operating at meaning level in an extremely determined way, following at least two paths: the first taking place

> «par des prototypes, qui attirent nos variations, comme d'autant d'attracteurs imaginaires, dont les traits se prêtent ensuite à la combinaison (Brandt 1994: 86–87)»; the second taking place «par des schèmes-standard qui s'appliquent à des classes entières de prototypes, et qui appellent, à titre de variation locale et minimale, commutative, autour d'un type, une réponse terminologique régulièrement positive». (Brandt 1994: 87)

In this case, there is a commutation toward the significant. In other words, in order to establish the existence of a morphology of content, it is necessary that continuous variation produce a discontinuous effect in the other plane every time. This fact is analogous to construction of expressive morphologies, but in an opposite sense. In speech situations there is a variation, or rather a deployment of the semantic domain, triggered by the variability of the parameters of the construction of meaning. The space of the form of content or the semantic morphology has to account these elements or dynamics as interfaces:

> Une morphologie du contenu est un «système» de positions ou d'états différents d'une même chose, un système défini par cette variation existante, que l'opération retrouve, cest-à-dire par le déploiement continu d'un domaine de sens que «recouvre» un ensemble discontinu de termes (expressions signifiantes). Une telle morphologie est en vérité une sémiologie, au sens de Barthes ou de Greimas; c'est une formation culturelle, consciente ou inconsciente, relevant d'une langue ou plus particulièrement d'un discours, et qui se manifeste par l'actualisation des présuppositions crées par ces contraintes pour ainsi dire projectives, reliant du signifiant discontinu (différentiel) a du signifié continu. Il s'agit ici de ce qu'on peut appeler la représentation symbolique du sens (Brandt 1994: 87–88).

Even for the morphologies of the content it is possible to build a modelization of their organization. Without going into the details of morphodynamical models, we try to translate this model into a more immediate language: let us suppose a space substrate W, whose segmentation in a system of thresholds or boundaries we aim to explain. It is a representation of what we call morphologies of the content, like cultural strata, operating both in language and in discursive practices:

> Dans sa version «élémentaire», la théorie morphodynamique considère W comme un espace «externe», qui «contrôle» des processus dynamiques définis par des fonctions potentiel fW: M àR. Autrement dit, W est l'espace «source» d'un champ σ W à F qui associe aux unités w de W des éléments fw de l'espace F des processus dynamiques choisis. Ces processus dynamiques déterminent alors les états internes d'un certain système S, et la catégorisation de W procédera de la qualification de ses unités w sur la base de l'identité qualitative (intrinsèque) des états internes que celles-ci contrôlent (Piotrowski 2005: 182).

The study of the morphology of the content enables us to enter into whole relations between matter, substance and form. By doing so, one can focus on the idea that the formal boundaries of languages emerge from a material that already has structuring internal trendlines (Eco 1997: 2002). This emergence organises forms as a spatial distribution of substances where condensations and stable or stabilized uses of linguistic types occur. Piotrowski, presenting his scheme, states that:

> La donnée d'une topologie et d'un groupe opérant sur F permet d'y introduire les notions de stabilité et de type qualitatif, et on montre que, sous certaines conditions, les formes dynamiques issues des stabilisations successives d'une dynamique instable f de F se distribuent dans F suivant une stratification KF d'interfaces (frontières) qui catégorise localement F. Les portions d'espace qui regroupent des formes de même type qualitatif sont mutuellement séparées par des frontières qui correspondent aux formes de degré d'instabilité supérieur. La catégorisation Kw de l'espace substrat W est alors «la trace sur lui des instabilités des formes [dynamiques] qu'il contrôle» et des rapports de conflit que les états internes entretiennent (Piotrowski 2005: 182).

What Hjelmslev identifies by matter is the «space of dynamics»: an area of instability, continuous but structured, coinciding with the experience as speakable. Thus, morphological interface of content does not constitute a projection of readymade elements on the space of matter, but it is the emergence of internal states that progressively become stable and construct thresholds and boundaries of discontinuities in the continuum of experience. Substance of content becomes the projection on the concrete surface of linguistic exchange and morphologies of content. Therefore, languages can be compared to these internal spaces of control.

6.7 Phenomenology of Sign: The Natural «Tranlatability» of Forms

Thus, we can conceive languages as places where there is a plurality of local content morphologies, which represent constraints for syntagmatic and discursive construction and for the social imaginary. The first problem emerging is the natural translatability of linguistic forms (the signs), intra or interlinguistic. In 13th paragraph of Prolegomena, Hjelmslev uses a really known example: the semantic domain of wood (bois in French). Starting from hjelmslevian example, one can realize what we mean when we say that the theme of semiolinguistic reflection is plasticity and dynamism of the semiolinguistic forms. These one are built on the idea of translatability, according to which the meaning, can be understood it is knowable because it is translatable:

This incongruence within one and the same zone of purports turns up everywhere. Compare also, for example, the following correspondences between Danish, German, and French:

Baum arbre

træ

Holz bois

skov Wald

forêt

We may conclude from this fact that in one of the two entities that are functives of the sign function, namely the content, the sign function institutes a form, the content form, which from the point of view of the purport is arbitrary and which can be explained only by the sign function and is obviously solidary with is. (Hjelmslev 1968: 54).

Under which conditions is continuous variation of the content possible? How to bring together the push differentiating languages and the similarities or homogeneities set by world? Why, even though speaking different languages, variation seems to be the same? For Hjelmslev, the area of matter about which we talk, is not merely a continuous amorphous, it has nothing smooth or inarticulate. Let us take, for example, the first border, i.e., Baum/Holz in German and arbre/bois in French. This is the same border in both paradigms: «elle indique sémantiquement un passage intuitivement net entre l'idée du végétal en question et l'idée de la matière fibreuse que l'on en tire, mais qui ne possède plus la forme du végétal. Le terme danois træ recouvre les deux idées, et ce n'est pas parce qu'un seul termes les rend, que les Danois les confondraient! Le jeu des articles dépend au contraire de cette distinction» (Brandt 1994: 93).

The second border, however, that Holz/Wald and bois/forêt, does not coincide in the two languages, and we deal with a quantitative difference: *bois* is larger then *Holz*, but, as an idea of set of trees, it is less large than *forêt*, *Wald*. In the paradigm of Danish language, this second border does not appear, so skov embraces the general idea of a land populated by trees; the frontier of the Danish træ/skov, projected on german and French paradigms cuts in two Holz and bois, marking a transition between the intelligible «fibrous material» and the «whole trees» because, as Brandt concludes, «ce sont les bois-Holz devenus skov qui vont voisiner forêt-Wald, et non pas les bois-Holz restant træ» (Brandt 1994: 94). In other words that concerns a semantic passage, hidden in the two paradigms, from the set of differential relations within the field that the analysis of Danish language makes manifest. As well as, conversely, the transition between «tree» and «fibrous material» is absent in Danish because of synthetic term. Brandt concludes:

> Il faut distinguer les sauts qualitatifs du sémantique et les frontières des paradigmes (la morphologie du signifiant ne recouvre pas la morphologie du signifié) dans une même langue; il faut projecter les articulations terminologiques de langue en langue pour dégager les articulations du contenu qui soutiennent les signes, et qui ne sont aucunement constituées par ces signifiants (qui tantôt les manifestent, tantôt les cachent), mais renvoient à une topologie sémantique rendant compte de ces «passages». Le domaine du sens hyléthique, pour ainsi dire, qui était le point de départ intuitif de Hjelmslev, et qui lui permettait de mettre en parallèles les paradigmes (…) est donc topologiquement structuré, et cette structure est bien «interlingue» (Brandt 1994: 94).

Hjelmslev intended to focus on the link between languages nature as systems in continuous differentiation, and semantic structures. If diversity is a universal trait of languages, the semantic structures constitute the bases, of a topological/dynamical nature, from which meaning can be extracted; and that because of their transverse movement with regard to different languages. Differential paradigms of languages do not obey to linear models of the organization, but are the linearization of underlying semantic topologies, ensuring the intrinsic translatability of the forms and meanings.

Let us go back to the example of the semantic field of the wood. Let us see how the linear model of Hjelmslev is inadequate, and how to integrate these lacks. So, we have seen that the field embraces three specific semantics zones—the vegetable, the material fiber and the ground—and includes at least two changes of state and a quantitative transition. Now, if we consider at first the continuous variation as a linear slip, it is possible to formulate a triangular representation able to explain the semantic components of the paradigm in question that, according to Brandt, could be understood as the "«image»" of sens, implied in expressive paradigms. With regard to German paradigm, we realize what we have said above. We are in front of three ideas involved (or three semantic areas): the set of trees, tree interpreted as vegetable, and the fibrous material obtained by cutting the plant. The German paradigm covers quite clearly the distinction between the three ideas except Holz. This is attracted by Wald when one inserts quantitative parameters about the magnitude and extension of the zone «set of trees»; concerning bois it seems to happen the same thing, while skov, as we have already said, is unique.

On the contrary, in relation to the idea of «vegetable», Baum and arbre are unique, and it is this uniqueness that Hjelmslev's graph does not allow to understand: if semantic proximity of Holz and Wald displays, as said, a quantitative parameter—a slip—as well as two semantics meanings of bois, the proximity of Holz and Baum, that is manifested, so to speak, by the two directions of træ, does not imply any quantitative parameter, but a sharp and categorical transition. In short, we cannot say that the three areas constitute an absolute semantic balance, from which properties we can derive semantic equivalences underlying paradigms: the same field exhibits situations of syncretism by quantification, sharp transitions or even by absence of syncretism. This is because, as Brandt writes, «les morphologies n'ont aucune raison d'être linéaires, et, dans notre cas, elles ne le sont certainement pas. La topologie qui sous-tend ces paradigmes, linéarisés par une routine sémiologique (…) doit en revanche saisir le contenu du contenu, les choses dont il s'agit dans le domaine de sens considéré» (Brandt 1994a: 95).

How to describe this implied topology in domain of sense? Hjelmslevian insights can be reconstructed, as long as one draw a nontriangular geometry, a cognitive and semantic script—the domain of potential meanings—that underlies the paradigmatic differentiation, namely the relationship between values. The description of a paradigmatic field goes through the identification of «semantic trajectories» that words take: evidences of discontinuity that experience, cognitive activity, and human cultural realize.

This is why, we have to abandon models of linear description of sense, and to think words in the terms of «trajectories»: «les états se trouvent ainsi reliés par des chemins, et se présentent comme des zones dans un paysage qui est celui de la catastrophe thomienne dite fronce ou cusp (Brandt 1994: 95)».

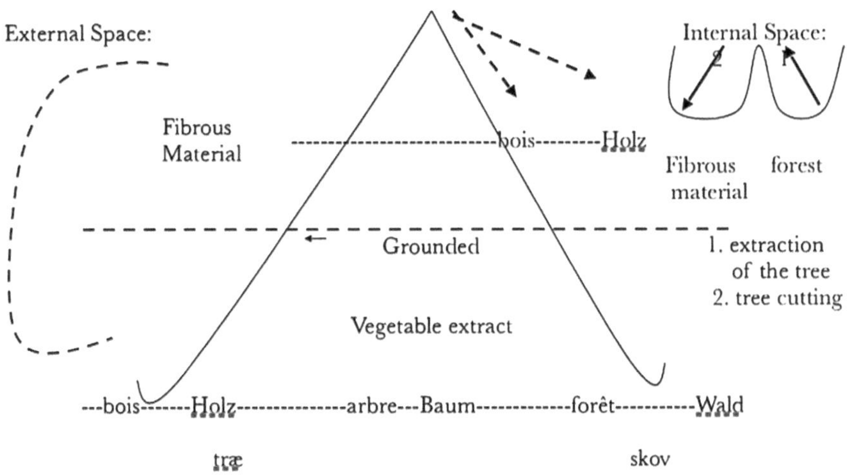

Source: Brandt 1994

This mental geography or topography of semantic building process binds these three ideas in a «set of vector trajectories» of meaning. This path is marked by two catastrophes: it is the presence of separation, because of two points of catastrophe, that allows to build these three linguistic paradigms. Hjelmslevian representation well grasps the differential dimension in interlinguistic relations. However, it is not able to describe the steps and slips of internal spaces making sure that a term like bois could simultaneously belong to the domain from which it is extracted, as well as to which one it has become, serving as an intermediary and agent of transformations.

Hjelmslev's proposal makes think sign dimension as a plastic movement of paradigms, constructing themselves through the same topology: it is this one that constitutes the semantic basis on which speakers translate paradigmatic nonidentities. This is why Brandt rethinks the most brilliant idea of Hjelmslev: the natural translatability of the semiotic dimension.

Thus, we can retrieve the genuine power of Hjelmslev's theory: think to the linguistic form not as a formal element to project on an amorphous material continuum, but as a «temporary situation», a «floating step», temporally and locally stabilized, between the continuity of the universe of speakable, its lines of internal organization and the concrete linguistic experiences. This idea of a «plastic sign» paves the way the study of the semiotic function as keystone to understand the stratification of language.

References

Ablali, D. (2003). *La sémiotique du texte: du continu au discontinu*. Paris: L'Harmattan.

Arrivé, M. (1986). «Y-a-t-il en glossématique une théorie de l'énonciation? *Histoire, Epistémologie, Langage, VIII*(2), 177–189.

Badir, S. (2000). *Hjelmslev*. Paris: Les Belles Lettres.

Badir, S. (2004). Immanence et cognitivisme. Un débat épistémologique autour de la sémantique cognitive. http://www.revue-texto.net/Inedits/Badir_immanence_html.

Badir, S. (2014). *Epistémologie sémiotique. La théorie du langage de Louis Hjelmslev*. Paris: Honoré Champion.

Barbaras, R. (2002). Le vivant comme fondement originaire de l'intentionnalité perceptive. In J. Petitot, F. Varela, B. Pachoud & J.-M. Roy (Eds.), (pp. 681–698).

Benveniste, E. (1966). *Problèmes de Linguistique Générale*. Paris: Gallimard.

Benveniste, E. (1974). *Problèmes de Linguistique Générale II*. Paris: Gallimard.

Bo, L.-J. (1997a). Sciences phonétiques et relations forme/substance. 1 Un siècle de ruptures, négociations et réorganisations. *Histoire, Epistémologie, Langage, XIX*(1), 5–41.

Bo, L.-J. (1997b). Sciences phonétiques et relations forme/substance: 2. Du poids de la substance sur la forme aux réarticulations scientifiques. *Histoire, Epistémologie, Langage, XIX*, 5–25.

Bondì, A. (2008). Regimi del senso. La teoria delle forme semantiche tra diversità linguistica e stabilizzazione semiotica. In F. Giuliani & M. Barni (Eds.), (pp. 239–252).

Bondì, A. (2009). La fonction sémiotique chez Hjelmslev. Du modèle structurale au modèle cognitif. *Histoire, Epistémologie, Langage*, *30*(2), 201–214.

Bondì, A. (2011). *La parola e i suoi strati. La semiotica dinamica di L. Hjelmslev*. Acireale-Roma: Bonanno.

Bourgine, P. & Lesne, A., (Eds.). (2006). *Morphogenèse. L'origine des formes*, Paris: Belin.

Brandt, P. A. (1990). Du bois. *Versus, 57*, 135–138.

Brandt, P. A. (1992). *La charpente modale du sens*. Philadelphia-Amsterdam: John Benjamins Publishing.

Brandt, P. A. (1993). De la linguistique structurelle à la linguistique cognitive avec Hjelmslev. *Travaux du Cercle Linguistique de Copenhague, XXIV*, 9–25.
Brandt, P. A. (1994). *Dynamiques du sens. Études de sémiotiques modale*. Aarhus: Aarhus University Press.
Brandt, P. A. (1995). *Morphologies of meaning*. Aarhus: Aarhus University Press.
Brandt, P. A. (2001). Le motif analytique des Prolégomènes (Omkring sprogteoriens grundlæggelse). In R. Galassi & M. De Michiel (Eds.), (pp. 101–114).
Brøndal, V. (1939). Linguistique Structurale. *Acta Linguistica. Revue Internationale de Linguistique Structurale*, Vol. 1, fascicule 1. Einar Munksgaard-Copenhague.
Bulea, C. (2005). Est-ce ainsi que les signes vivent?. http://www.revue-texto.net/Saussure/Sur_Saussure/Bulea_Signes.html.
Cadiot, P., & Visetti, Y.-M. (2001). *Pour une théorie des formes sémantiques. Motifs, profils, thèmes*. Paris: PUF.
Chomsky, N. (2000). *New horizons in the study of language and mind*. Cambridge: The Press Sybdicate of the Univeristy of Cambridge.
Coquet, J.-C. (2008). *Phusis et langage. Une phénoménologie du langage*. Presses Universitaires de Valenciennes.
Culioli, A. (1990). *Pour une théorie de l'énonciation. Opérations et représentations. Tome 1*. Paris: Ophrys.
Culioli, A. (1999a). *Pour une théorie de l'énonciation. Formalisation et opérations de repérage. Tome 2*. Paris: Ophrys.
Culioli, A. (1999b). *Pour une théorie de l'énonciation. Domaine notionnel. Tome 3*. Paris: Ophrys.
Culioli, A., & Normand, C. (2005). *Onze rencontres sur le langage*. Paris: Ophrys.
Deleuze, G., & Guattari, F. (1980). *Mille plateux. Capitalisme et schizophrénie*. Les éditions de Minuit: Paris.
de Saussure, F. (1922 [1995]). *Cours de Linguistique Générale*. Paris: Payot.
de Saussure, F. (2002). *Ecrits de linguistique générale*. Paris: Gallimard.
Eco, U. (1997). *Kant e l'ornitorinco*. Milano: Bompiani.
Fontanille, J. (2004). *Soma et Séma. Figures du Corps*. Paris: Maisonneuve et Larose.
Franckel, J.-J. (1998). Référence, référenciation e valeurs référentielles. *Sémiotiques, 15*, 61–84.
Givon, T. (2006). Bio-linguistics. *The Santa Barbara Lectures*. Philadelphia: John Benjamins Publishing.
Harder, P. (2001). Form, substance and function: the importance of structure in a functional lingustics. In R. Galassi, & M. De Michiel (Eds.), (pp. 9–24).
Hjelmslev, L. (1938a). Essai d'une théorie de morphèmes. Actes du IV Congrès International des Linguistes, Kopenhavn, 1936. (pp. 140–151). In L. Hjelmslev (1971), pp. 161–173.
Hjelmslev, L. (1938b). Ueber die Beziehungen des Phonetik zur Sprachwissenschaft. *Archiv fur vergleichende Phonetik, II*, 129–134.
Hjelmslev, L. (1939a). Notes sur les oppositions supprimables. *TCLC, 8*, 51–57. In L. Hjelmslev (1971), pp. 91–97.
Hjelmslev, L. (1939b). A causerie on linguistic theory. *TCLC, XIV*(1973), 101–118. In L. Hjelmslev (1985), pp. 69–76.
Hjelmslev, L. (1939c) The syllable as a structural unit. *Proceedings of the third international congress of phonetic sciences, Ghent* 1938, (pp. 266–272). In L. Hjelmslev (1985), pp. 165–174.
Hjelmslev, L. (1939d). La structure morphologique. *V Congrès International des Linguistes, Bruxelles 1939,* (pp. 66–93). In L. Hjelmslev (1971), pp. 122–147.
Hjelmslev, L. (1939e). Le caractère linéaire du signifiant. *V Congrès International des Linguistes, Bruxelles 1939*. In Zinna (1995), pp. 249–257.
Hjelmslev, L. (1939f). La notionde rection. *Acta Linguistica, 1*, 10–23. In L. Hjelmslev 1971, pp. 148–160.
Hjelmslev, L. (1943a). Langue et parole. *CFS, 1*. In L. Hjelmslev (1971), pp. 78–90.
Hjelmslev, L. (1943b). Omkring sprogteoriens grundæggelse. *TCLC, XXV*(1993), 3–141.
Hjelmslev, L. (1947). The basic structure of language. *TCLC, XIV*(1973), 119–156.

Hjelmslev, L. (1948a). Linguistique structurale. *Acta Linguistica, IV*, V–XI. In L. Hjelmslev (1971), pp. 29–34.

Hjelmslev, L. (1948b). Structural analysis of language. *Studia Linguistica, I*, 69–78. In L. Hjelmslev (1971), pp. 35–44.

Hjelmslev, L. (1950). Le rôle structural de l'ordre des mots. *Journal de Psychologie normale et pathologique, 43*, 54–58.

Hjelmslev, L. (1953). Sorigets indholdsform som samfundsfaktor. *Det danske Magasin, II*, 1–7. In L. Hjelmslev (1971), pp. 98–104.

Hjelmslev, L. (1954). La stratification du langage. *Word, 10*, 163–188. In L. Hjelmslev (1971), pp. 45–77.

Hjelmslev, L. (1956a). Animet inanim personnel et non-personnel. *Travaux de l'Institut de Linguistique (Paris), I*, 155–199. In L. Hjelmslev (1971), pp. 220–259.

Hjelmslev, L. (1956b). Numerus et genus. *Janus, III*, 11–24.

Hjelmslev, L. (1957). Pour une sémantique structurale. *Report for the english international congress of linguists on the question: Dans quelle mesure les significations des mots peuvent elles être considérées comme formant une structure? Oslo, 1057*, 636–654. In L. Hjelmslev (1971), pp. 105–121.

Hjelmslev, L. (1961). *Prolegomena to a theory of language*. Wisconsin: The Regents of the University of Wisconsin.

Hjelmslev, L. (1962). Some reflexions on practice and theory in structural semantics. *Language and society, essays presented to Arthur M. Jensen*, 55–63.

Hjelmslev, L. (1963) *Sprøget. En introduktion, Charlottenlund*. The Natur Method Center.

Hjelmslev, L. (1971). *Essais Linguistiques*. Paris: Les éditions de Minuit.

Hjelmslev, L. (1975). Resumé of theory of language. *TCLC, XVII*, 1–279.

Hjelmslev, L. (1985). *Nouveaux Essais*. Paris: Les éditions de Minuit.

Lassègue, J. (2003a). Remarques anthropologiques sur la question de l'émergence du sens. In Siksou (Ed.), pp. 321–335.

Lazard, G. (2006). *La quête des invariants interlangues*. Paris: Honoré Champion.

Lazard, G. (2007). La linguistique cognitive n'existe pas. *Bulletin de la Société Linguistique de Paris, 102*(1), 3–16.

Lo Piparo, F. (1991). Le signe linguistique est-il à deux faces? Saussure et la topologie. *Cahiers Ferdinand de Saussure, 45*, 213–221.

Maniglier, P. (2006). *La vie énigmatique des signes. Saussure et la naissance su structuralisme*. Paris: Léo Scheer.

Maturana, H., & Varela, F. (1980). *Autopoiesis and cognition. The realization of the living*. Dordrecht: Reidel Publishing Company.

Ouellet, P. (1994). La sémiotique cognitive: les sciences de l'esprit entre la nature et la culture. *Sémiotiques, 6–7*, 137–159.

Petitot, J. (1985). *Morphogenèse du sens*. Paris: PUF.

Petitot, J. (1989). Hypothèse localiste, modèles morphodynamiques et théories cognitives: remarques sur une note de 1975. *Semiotica, 77*, 65–109.

Petitot, J. (1991). Syntaxe topologique et grammaires cognitives. *Langages, 103*, 97–128.

Petitot, J. (1996). La généalogie morphologique du structuralisme. *Critique*, 620–621. *Numéro spécial en hommage à Claude Lévi-Strauss*, 97–122.

Petitot, J. (2002). Le nervature del marmo. Osservazioni sullo zoccolo duro dell'essere in Umberto Eco. In J. Petitot & P. Fabbri (Eds.), pp. 71–92.

Petitot, J., & Fabbri, P. (Eds.). (2002). *Nel nome del senso. Intorno all'opera di Umberto Eco*. Milano: Sansoni.

Petitot, J., Varela, F., Pachoud, B., & Roy, J.-M. (Eds.). (2002). *Naturaliser la Phénoménologie. Essais sur la phénoménologie contemporaine et les sciences cognitives*. Paris: Les éditions du CNRS.

Pétroff, A. (1995). L'ordre et désordre: l'interaction Langue-Parole. In M. Arrivé & C. Normand (Eds.), pp. 369–386.

Pétroff, A. (2004). *Saussure: la langue, l'ordre et le désordre*. Paris: L'Harmattan.

Piotrowski, D. (1993). Sur le statut du concept de matière. *TCLC*, *XXIV*, 89–111.
Piotrowski, D. (1994). Approche morphodynamique de deux régimes fondamentaux du langage. *Sémiotiques, 6–7*, 175–186.
Piotrowski, D. (1997). *Dynamiques et structures en langue*. Paris: Les éditions du CNRS.
Piotrowski, D. (2005). *Du sens en langue, Cahiers du CREA, 21*. Paris: CREA/Ecole Polytechnique.
Piotrowski, D. (2009). *Formes et objectivités linguistiques*. Paris: Honoré-Champion.
Porte, M. (Ed.). (1994). *Passion des Formes. Dynamique qualitative, sémiophysique et intelligibilité. À René Thom*. Fontenay-St. Cloud, ENS éditions.
Rastier, F. (2001a). Du signe aux plans du langage ou de Saussure à Hjelmslev. In R. Galassi & M. De Michiel (Eds.), 161–181.
Rastier, F. (2001b). *Arts et sciences du texte*. Paris: PUF.
Rastier, F. (2006). *La structure en question* (pp. 93–104). VI: Janus.
Salanskis, J.-M. (2003). *Heménéutique et cognition*. Lille: Presses Universitaires de Septentrion.
Sebeok, Th, & Umiker-Sebeok, J. (Eds.). (1992). *Biosemiotics: The Semiotic Web*. Berlin: Mouton de Gruyter.
Siblot, P. (Ed.). (1997). Langue, Praxis et production de sens. *Langages*, *127*, 3–8.
Siksou, J. (2003). *Variation, construction et instrumentation du sens*. Paris: Lavoisier.
Stjernfelt, F. (1992). Categorical perception as a general prerequisite to the formation of signs? *On the biological range of a deep semiotic problem in Hjelmslev's as well as Peirce to semiotics*. In Th. Sebeok & J. Umiker-Sebeok (Eds.), pp. 427–454.
Thom, R. (1986). *Esquisse d'une sémiophysique*. Paris: InterEditions.
Victorri, B. (2005). Continu et discret en sémantique lexicale. *Cahiers de Praxématique, 42*, 75–94.
Visetti, Y.-M., & Cadiot, P. (2006). *Motifs et proverbes. Essais de sémantique proverbiale*. Paris: PUF.
Wildgen, W. (1993). Hjelmslev's treatment of the substance of content and the penury of linguistic structuralism. *Hjelmslev colloquim in Copenhagen*, 25–27, November 1993.
Wildgen, W. (2005b) *Semiosis and Catastrophes. René Thom and Semiotic Heritage*. Centro Semiotico di Urbino.
Zinna, A. (2008). Il primato dell'immanenza nella semiotica strutturale. Relazione presentat al convegno. Incidenti ed esplosioni. In A. J. Greimas e & J. M. Lotman (Eds.), *Per una semiotica delle culture*. www.ec-aiss.it/archivio/tipologico/autore.php.

Chapter 7
From Topology to Quasi-Topology. The Complexity of the Notional Domain

Francesco La Mantia

7.1 Notional Domains

Notional domains are a privileged metalinguistic construction of the theory of predicative and enunciative operations (henceforth, TOPE). As Filippi-Deswelle (2008: 301) states: «Le domaine notionnel d'Antoine Culioli est [...] un outil de représentation metalinguistique [...]». They are introduced in the context of formal analysis, and represent the lexical units as *markers*—i.e. as «traces» of language operations underlying the modes of utterances formation. (See at least Culioli 2002 and Culioli 2014). The metalinguistic use of notional domains involves mainly two language operations: the *fragmentation of the notion* and the *construction of existence of abstract occurrences*.

7.1.1 Notions

First of all, we must clarify what the notions for TOPE are. They are therefore structured sets (or *bundles*) of properties that include specific aspects of biological, physical and cultural life of human beings. Each of these properties is the *cognitive indicator* of interactions finely distributed between individuals and individuals, individuals and the environment, individuals and artefacts of different kinds: «A notion can be defined as a complex bundle of structured physic-cultural properties. [...] Notions involve properties [...] coming from interaction between people and

F. La Mantia (✉)
Dipartimento Di Scienze Umanistiche, Università Degli Studi Di Palermo, Palermo, Italy
e-mail: francesco.lamantia28@unipa.it

F. La Mantia et al. (eds.), *Language in Complexity*,
Lecture Notes in Morphogenesis, DOI 10.1007/978-3-319-29483-4_7

people, people and objects, from biological constraints, from technical activitity, etc.»[1] Using a different terminology: notions correspond to cores of heterogeneous experiences in which linguistic and nonlinguistic aspects of human cognition are mixed together. From a formal point of view, each notion is identifiable as an unsaturated predicative entity such as «to be *x*». For instance, «to be fox».

This predicate is the metalinguistic counterpart of the notion /fox/, which is a set of properties that includes the main symbolic and experiential contents associated with the idea of fox or (in french) «vulpeité»: «[…] la notion va ressembler tout un ensemble de propriétés que vous, vous allez ramener […] à la vulpeité […] Bref, à «être rénard»».[2] As for the fragmentation and the construction of existence, they are, roughly, language operations that allow passing from notions to utterances (or textual portions) which manifest them.

7.1.2 Fragmentation and Constructions of Existence

In particular, fragmentation is a language operation assigning a notion the corresponding extension—i.e., the set of abstract occurrences (or possible and enumerable samples) that instantiate it. As Culioli (1989: 181) affirms: «This operation makes the notional representation fragmentable, that is, endowed with extensionality. We thus construct an aggregate of abstract occurrences that possesses the notion in question […]. The Quantifiabilization (i.e. fragmentation) is a mean to construct the extension of a notion». Each of these occurrences is a cognitive artefact that corresponds, roughly, to a possible combination of the notion's properties. In other words: they are virtual examples built from assemblies of properties that make different presentation modes of the same notion.

Then the fragmentation of /fox/ is the language operation that builds abstract occurrences of /fox/—i.e., the possible modes of presentation of such notion. In short: a fox «so and so» a fox «so and so» a fox «so and so», etc. According to Culioli (2002: 125): «J'ai le prédicat «être chien». À partir de ça, je vais construire […] l'agrégat (si l'on voulait parler en termes de logique, on dirait la «classe» […]) des occurrences abstraites de «être chien»: un chien+un chien+un chien, etc. […]». The construction of existence is the next step after fragmentation. It is a language operation that provides the verbal formulation (or the *mise en forme*) of abstract occurrences, namely the creation of a family of utterances through which the occurrences are somehow utterable. For instance: the *mise en forme* of the abstract occurrences of /fox/ is given by the constitution of a family of utterances such as «That's really a fox»!; «No, it's absolutely not a fox!»; «It's hardly a fox»! «It's a fox, a real fox!»; and so on.

---

[1]Culioli (2000a, b, c, p. 69).

[2]Culioli (2002, p. 53).

In other words: if the fragmentation is an operation of language which builds the abstract occurrences of a notion, the construction of existence is a language operation which *localizes* each of these occurrences in an actual situation of discourse—or, in the terms of TOPE, which transforms abstract occurrences of the notion into *situated* enunciative occurrences.

Hence a complex chain of operations which takes the name of «quantification»: «La quantification permet [...] d'effectuer deux opérations essentielles: (1) la quantifiabilitsation (ou, pour ce qui préfèrent un terme plus simple, la fragmentation). [...] (2) [...] construire l'esistence d'une occurrence [...] en la situant dans l'espace-temps énonciatif».[3] Notional domains are the metalinguistic constructs that allow describing the results of these operations. In particular, they are used to describe situated occurrences of a notion. According to Gerner (1995: 39): «Culioli développe l'idée d'un domaine [...] associé à chaque notion dégagée dans une chaîne parlée».

7.2 Notional Domains and Intuitive Topologies

A notional domain is the metalinguistic counterpart of a class of abstract occurrences: «We (that is, linguists as much as they attempt to represent metalinguistically the activity of subjects) are thus led to construct a notional domain which structures the class of occurrences of a given notion».[4] In more detail: a notional domain is the class of abstract occurrences of a notion that *instantiates* the interior, the exterior and the boundary of a given set—i.e. it is a collection of possible notional occurrences that can be defined in terms of an intuitive elementary topology. As Gerner (1995: 39) affirms: «Culioli s' inspire de la topologie qui est un domaine des mathèmatiques pour représenter ce concept du domaine notionnel». The reference to the conceptual repertoire of this intuitive topology must be adequately justified. So it will be clarified: (a) what interior, exterior and boundary of a given set are; (b) how a class of abstract occurrences can be defined in terms of these concepts; (c) how the notional domains allow describing the situated occurrences of a notion.

7.2.1 *Interior, Exterior, Boundary: From the Basic Concepts to the Definitions*

In the technical terminology of the general topology (Janich 1994), «interior», «exterior» and «boundary» are used in reference to sets of points (or elements) endowed with peculiar properties. The exam of these properties is based on the concept of topological space and the local concepts of interior point, exterior point and boundary point of a given set. First, we need to clarify each of these concepts.

[3]Culioli (2000c, pp. 83–84).

[4]Culioli (2000a, p. 70).

7.2.1.1 Topological Spaces

A topological space is a structure constituted by a non-empty set where a class of non-empty sub-sets – called "neighbourhoods" – is assigned to each point of it so to satisfy certain axioms. We labels with *X* the non-empty set and with $U = \{U_x\}$ $x \in X$ the «family» of classes of non-empty subsets of *X*—or neighbourhoods of *x*. The axioms in question specify the fundamental relationship between the neighbourhoods of each point of *X* and they define a topology on *X*. In particular, they affirm that:

(a_1) every neighbourhood of *x* contains *x* and *X* is the neighbourhood of each of its points;
(a_2) the subset of a neighbourhood which contains *x* is also a neighbourhood of *x*;
(a_3) the intersection of two neighbourhoods of *x* is a neighbourhood of *x*;
(a_4) every neighbourhood of *x* that contains a neighbourhood of *x* is also a neighbourhood of each of its points.

We will say that *U* defines a topology on *X* if and only if each member of *U* satisfies the axioms (a_1), (a_2), (a_3), (a_4). A topological space can be so defined as a pair of elements (*X*, *U*)—where *X* is a non-empty set and *U* a family of classes (or sets) of non-empty subsets of *X* for which established axioms will remain valid. Unless it is strictly necessary, the reference to the topology *U* is removed from the notation, which is thus reduced to the single letter *X*.

7.2.1.2 Interior, Exterior and Boundary Points

The concept of interior point, exterior point and boundary point has been introduced from the concept of topological space. Given a topological space *X* and a non-empty subset *A* of *X*, a point *x* of *X* is said interior, exterior or boundary point of *A* if, respectively, *A*, the complement of A with respect to *X* or none of two is a neighbourhood of *x*. All this can be displayed graphically.

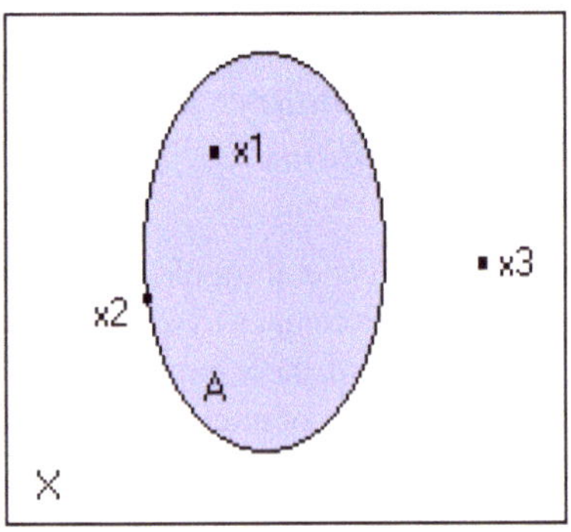

The subscribed numerical indices allow to distinguish adequately each type of point: x_1 is an interior point of A, x_3 is an exterior point of A (i.e. is an interior point of the complement of A with respect to X) and x_2 is a boundary point of A. Hence, the usual definition of interior, exterior and boundary of a given set.

7.2.1.3 Interior, Exterior and Boundary

So: it is called interior of *A* the set of interior points of *A*—and it is labeled with *int(A)*. They are called exterior and boundary of *A*, respectively, the interior of the complement of A with respect to *X*, labeled with *ext(A)* or *int(X\A)*, and the set of points of *X* belonging neither to the interior nor to the exterior of *A*—labeled with *B(A)*. As stated by Lipschutz (1969: 69–70): «Let *A* be a subset of a topological space *X*. [...] The set of interior points of *A* [...] is called the interior of *A* [...] the exterior of *A* is the interior of the complement of *A* [...], the boundary of *A* is the set of points which not belong to the interior or the exterior of *A*». Here's another diagram:

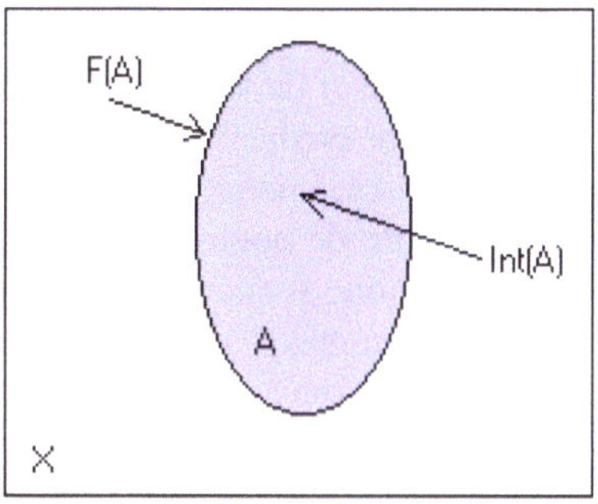

Let us move now on to the second point.

7.3 Metalinguistic Applications

The metalinguistic analysis of fragmented notions is centred on the topological concepts of interior, exterior and boundary of a given set. TOPE employs such concepts to describe a particular aspect of the classes of abstract occurrences—namely the fact that each of them provides a «boundary».

Metaphors aside: for each constructed class it is possible to distinguish between occurrences that are members of the class, occurrences which are not at all and others which are (or are not) only to a certain extent, even small. The topological concept of interior is used for all the first type occurrences. The topological concepts of external and boundary are used, respectively, for all the second and the third type occurrences. Hence the partition (well established in the lexicon of TOPE) among interior, exterior and boundary of a notional domain: «[...] définir un domaine notionnel revient à envisager une classe d'occurrences d'une notion (donc la rendre quantifiable) et, d'un point de vue qualitatif, construire un espace topologique qui permettra de distinguer ce qui appartient au domaine (l'intérieur) [...] ce qui n'a pas vraiment les propriétés réquises (la frontière) ce qui est autre (l'extérieur)».[5] This partition can be justified on the basis of some in-depth analysis of the notion.

7.3.1 Notional Networks, Polar Pairs and Abstract Spaces

These are essentially three insights. The first twos concern the metalinguistic representation of notions: each notion is described as the *node* of a network of (similar or dissimilar) notions and to each node are associated two predicates. For example: the notion /fox/ can be described, *inter alia,* as a node of the notional network / Canid/ which is associated to the pair of predicates, «to be fox»/«not-to be fox». Generalizing: each notion is the node of (at least) a notional network described by a pair of predicates—or «polar pair of predicables» as in «to be P»/«to be not-P».

As Desclés (1998: 1) affirms: «Une notion /P/ peut être conçue comme étant définie à partir d'une couple polaire de notions prédicatives [...]. Une notion est intégrée dans un réseau d'autres notions (notions voisines, notions associées par une activité discursive, notions contraires...)». The same author, elsewhere, states more cautiously: «La notion (dans un sens proche de celui de Culioli) [...] peut être vue comme étant définie à partir d'une couple polaire de prédicables».[6] As for the third insight, it is relative to the topological coding of the abstract occurrences of a fragmented notion: by means of this coding, abstract occurrences are described in terms of points of a mathematical space associated with the notion and equipped with an intuitive elementary topology that provides an interior, an exterior and a boundary. An important consequence of this encoding is that TOPE uses the words «point» and «occurrence» as equivalent: «I use the term «point» as equivalent to «occurrence»».[7] The two steps—metalinguistic representation of the notions as nodes of notional network associated with polar pairs of predicable and topological coding of occurrences as points of a given abstract space—are fundamental preconditions to build up the notional domain. Let us see why.

[5]Bouscaren et al. (1987, p. 146).

[6]Desclés (1997, p. 14).

[7]Culioli (2000a, b, c, p. 71).

7.3.2 *Otherness and …*

The first step records a crucial aspect of notions at metalinguistic level. i.e. they are integrated in notional networks (or categories) that provide, for each, the identification of peculiar properties and the exclusion of the properties «perceived as others».[8] In brief, for every notion /P/, the members of the pair «to be-P»/«to be not-P» respectively encodes the constitutive properties of /P/ and properties that differ from the properties of /P/—where the rejection of the «other» properties is the specular counterpart of peculiar properties. As Vignaux (1988: 123) states: «Toute identification de propriété a son «complémentaire»: ce sont les propriétés qu'en «miroir», on pourra dire «autres»».

The network constraints specify these properties. Exemplifying (and simplifying): the identification of peculiar properties of /fox/ is subjected to the constraints of the /Canid/ category and is based on the rejection of other properties that are owned by the entire category—i.e. constitutive properties of different notions from /fox/ but nonetheless integrated into /Canid/. The class of abstract occurrences of a notion is so equipped with boundary in the sense that starting from the constraints of some notional category it is possible to construct both occurrences which possess only constitutive properties of the notions (and which fall under «to be P») both occurrences which possess different properties (and which fall only under «to be not-P»). Above all: hybrid occurrences, namely mixtures of peculiar properties and other properties that fall under the predicate «to be more or less P». This predicate, in conjunction with the previous, outlines all the possible articulations of a fragmented notion.

7.3.3 *… and Notional Spaces*

As for the second step, it allows to encode possible articulations of a fragmented notion at topological level and then construct the notional domain associated. In particular: for every notion /P/, abstract occurrences that fall under «to be P» instantiate the points of an open set *A*—where every open set is a collection that is the neighbourhood of its points and that coincides with its interior: «un ensemble est ouvert lorsqu'il coïncide avec son propre intérieur».[9] The choice of such metalinguistic representation is justified by the fact that, at this level of analysis the topological coding makes abstraction from the boundaries that «enclose» the class of abstract occurrences of /P/ and they separate it from classes of occurrences of other notions provided by the category. In the simplest case: from the classes of occurrences of another notion—henceforth, /Q/.

[8]Dufour (2011, p. 3).

[9]Desclés (2002, p. 249).

As affirmed by Culioli (1986: 87): «Nous prenons l'ouverture, c'est à dire que nous ne introduisons pas de propriété différentielle». A set is open if and only if it is disjoint from its boundary, then it is an area which provides «no first point, no last point»,[10] and is therefore an ideal candidate to describe aggregates of units considered exclusively from the interior, or, metaphor aside, ignoring what does not belong to these aggregates: «[...] en prenant l'ouverture du domaine, on n'a pas introduit de coupure qui ferait que vous auriez une où on dit «cela a telle propriété» et d'un autre côte «cela n'a plus telle propriété»».[11] In more detail: seen as points of A, the occurrences of /P/ are provided with one neighbourhood that is wholly contained in A: «On construit la classe des occurrences p_i, p_j di p. Toute occurrence p_i de p a, par construction, un voisinage et il existe donc une autre occurrence p_j de p. On voit donc que p est représentable par un ouvert».[12]Moreover, the points of each of these neighbourhoods correspond to occurrences of /P/ identifiable with each other, i.e. recognizable as exemplars of the same notions: «Ainsi, par voisinage, toute occurrence est identifiable à toute autre occurrence».[13]

By extension from local to global: all occurrences that instantiate the points of neighbourhoods entirely contained in *A* are recognizable as representative sample of /P/. Conversely: abstract occurrences that fall under «to be not-P» instantiate internal points of the complement of *A*. Hence, an in-depth analysis of the topological coding which introduces, in addition to an open set, its complement: «On construit alors le complémentaire C_p de cet ouvert».[14] Even this metalinguistic representation is justified: the complement of *A*, roughly, is anything not contained in *A* and abstract occurrences which instantiate its internal points are radically different from occurrences of /P/—i.e. the occurrences of /Q/. However, it is the case that the complement of an open set is a closed set, i.e. a set that is the union of an interior and a boundary: «C_p est, par définition fermé (il comprend la frontière e l'intérieur de Cp)».[15] Topological encoding of abstract occurrences of /Q/ does not make thus abstraction from the boundaries that separate them from the abstract occurrences of /P/. The next step is specular: we move from *A* to the closure of *A*, i.e. a set equipped with boundary that includes *A* as its own interior and which guarantees the description of the boundaries that separate abstract occurrences of /P/ from the abstract occurrences of /Q/: «nous construisons alors le fermé de l'intérieur de C_p».[16] TOPE calls «interior» *A* and «exterior» the interior of the complement of *A*. Along with the boundary of *A*, they constitute the notional domain of /P/. Moreover: the interior of the notional domain is «centred», i.e.

[10]Culioli (2000a, p. 70).

[11]Ibidem, p. 87.

[12]Culioli (1978a, b, p. 130).

[13]Culioli (2000a, p. 95).

[14]Culioli (1978b, p. 140).

[15]Ivi.

[16]Ivi.

equipped with a central point that corresponds to the typical occurrence (or *organizing centre*) of the notion and with which are identified all occurrences that instantiate internal points of A: «[...] l'intérieur—ensemble ouvert regroupant toutes les occurrences [...] identifiables à un type incarnant la notion».[17] The distribution of occurrences on the domain shall be fixed in accordance with the variable degrees (or gradients) of conformity that they exhibit with respect to the organizing centre. This, briefly, is the intuitive topology of a notional domain, or, in the lexicon of Gerner (1995: 40), the structure of a «notional space». It is to understand to what extent such intuitive topology conforms to the conceptual repertoire of general topology.

7.4 Some Difference: The Boundary as a Case Study

The literature of reference that has dealt with this problem, as Gerner (1995), Desclés (2002, 2012), guarantees the reliability of TOPE about the correct metalinguistic use of the examined topological concepts. On the other hand, Grize in Culioli (1981: 61–62, 1978a: 71) seem to suggest that the concept of notional domain is based on an accurate yet quite free (and therefore «unorthodox») of the available topological repertoire. An example in this direction is offered by the culiolian use of the concept of boundary. Let us see why.

7.4.1 From the Boundary as Limit ...

If compared to a classic topological approach, TOPE employs the concept of boundary in a relatively atypical way. The main indicator of such differences is given by the fact that, in this first context of discourse, the word «boundary» is used in reference to intervals with a certain breadth (or thickness, in Desclés' terms 2012), while in topology the same word designates the formal entity exemplified by threadlike lines or, in the most general case, by infinitely thin edges, i.e. by «lieu(x) sans épaisseur».[18]

With a more intuitive terminology, TOPE sees the boundaries as intervals with a certain extension. In the Mellet's words (Mellet 2009a: 11): «[...] la frontière notionnelle est d'épaisseur non-nulle. Ce n'est pas donc une ligne [...] mais une intervalle qui offre une certaine étendue». Hence the use of metalinguistic representations different in many ways from the ones based on the idea of the boundary as defined in the context of general topology. The reasons for this gap are all in the type of (notional) boundary chosen by TOPE: viewed as intervals with a certain

[17]Mellet and Monte (2009, p. 42).

[18]Desclés (2012, p. 12).

width, i.e. as «interstitial spaces» between the interior and the exterior of the domain, they are different from topological boundaries that appear as «extensionless» objects, or, in the Desclés' words (2002: 250) as «simples limites sans épaisseur». It must be said therefore that the types of notional boundaries processed by TOPE are more than one—and among them, some are compatible with the properties of classical topological boundaries. The notional boundaries can take both the form of intervals with a certain extension and the form of points without thickness—or, in the two-dimensional case, lines of points. As Ducard (2009: 64) observes: «[…] une frontière […] peut être une zone de transition […] ou peut se ramener […] à une limite ou à un bord». The first type of notional boundaries (i.e. intervals with a certain extension) is that one preferred by TOPE.

The use of this particular type depends on the fact that it ensures a considerable qualitative refinement of metalinguistic representations associated to the classes of abstract occurrences of each notion. The level of detail of representations based on classical topological boundaries is in fact lower than the level of detail of the ones based on boundaries seen as zones (or *interstitial spaces*). The difference of granularity can be grasped by comparing the encoding types that the two kinds of representation process for hybrid notional occurrences. In one case, it is a very poor coding: hybrid occurrences are described, at most, in terms of «boundary points» (or «oscillation points»—cf. Mellet 2009a, b), between the interior and exterior of the domain. In the other, the encoding is much subtle: the occurrences are distinct in reference to the number of properties that they share with the notion and are described in terms of points «situated» near the interior or in terms of points «situated» near the exterior (cf. Mellet and Monte 2009).

The difference of granularity is therefore a consequence of the different types of structure that are implemented by each representation. If the implemented structure is a topological structure, the level of detail of representations will be minimal: we will have to do with formal devices that provide clear lines of separation between interior and exterior of each domain—where each line is a special case of topological boundary. If the implemented structure provides interstitial spaces between the two kinds of «specular opens», the detail level of the representations will be more deep: we will have to do with formal systems that make possible taking account of the differences between the various types of hybrid occurrences. It remains to identify then the mathematical form of the structures which provide this kind of metalinguistic specifications.

7.4.2 … to the Boundary as Zone

According to Desclés (1998: 8), they are «quasi-topological (or locological)» structures (cf. De Glas 1990)[19]: «Cette notion est intégrée […] dans […] un espace

[19]See at least De Glas and Plane (2005), De Glas (2009).

abstrait (éventuellement muni d'une structure de quasi-topologie ou mieux d'une structure de locologie au sens de Michel De Glas)». The mathematics of locological spaces describes adequately the idea of «boundary as zone». Locological spaces are in fact mathematical entities with peculiar properties. So it is necessary to examine this notion and to understand the role it plays (or could play) in TOPE.

7.5 Locological Spaces: Definition and Basic Concepts

First, it should be explained what a locological space is. I will not examine all the aspects of the concept, and I will not go into the analysis of foundational issues. The inquiry will rather take the form of a preliminary specification of some locological elementary concepts. For this purpose, I will work on a definition of locology introduced in De Glas and Desclés (1996). According to my knowledge, it is one of the simpler definitions in literature. I will try to expose the contents—and, if necessary, to integrate them in a wider framework. I will intend, therefore, to refer to more recent works, including De Glas and Plane (2005) or Breysse and De Glas (2007); not before, however, discussing the preliminary concepts.

7.5.1 Universes and Binary Relations

According to De Glas and Desclés' definition (1996), the expression «locological space» is used to denote an abstract structure consisting of a non-empty set, called «universe», and a binary relation on the universe that satisfies three properties. To see which properties they are, we introduce some letters in place of the main used terms. In particular, we indicate with X the universe and with λ the binary relation on X. We will say that λ defines a locology on X if and only if:

(1) for all $x \in X$, $x \lambda x$, i.e. λ is reflexive;
(2) for all $x \in X$, $\lambda [x] \supsetneq \{x\}$, i.e. λ is «not thin»[20]—where $\lambda [x]$ is a non-empty subset of the universe (i.e. $\lambda [x] \subset X$) and denotes the set of the all images of x under λ, thus: $\lambda [x] = \{y: x \lambda y\}$;
(3) there exists a $\lambda' \subset \lambda$ that verifies (1) and (2)—and which is, inter alia, a symmetric relation. To sum up: λ corresponds to a locology on X if and only if it is reflexive, «not thin» and such to contain a binary relation λ' which is symmetrical (also reflexive and not thin). The pair (X, λ) represents a locological space: «Soient X un ensemble e λ une relation sur X.[…] Une relation λ

[20]see De Glas and Desclés (1996, p. 30).

> su X telle que (i) λ est réflexive, (ii) λ est non maigre, (iii) λ contient une relation symétrique satisfaisante (i) et (ii) est appelée une *locologie* sur X. Le couple (X, λ) est appelé un *espace locologique*[…]».[21]

Each of these properties must be examined in the light of the intuitive meanings assigned to λ—and, in particular, from the fact that every locology is thought as the formal counterpart of the binary predicate "(…)" *to be close to* "(…)" or "(…)" *to be the neighbour of* "(…)". It can be sketched a first interpretation of λ as proximity (or nearness) relation on X[22] and read the three previous conditions in terms of this relation. So:

(1) λ is reflective in the sense that every point x of the universe on which the locology is defined is close to itself (or is the neighbour of itself);
(2) λ is «not thin» in the sense that the set of all points near to (or «neighbours» of) any point x of the universe on which the locology is defined is not reducible to the singleton $\{x\}$;
(3) λ contains a reflexive relationship λ′ not thin and symmetrical in the sense that the universe on which the locology is defined admits at least a subset K which satisfies the first two properties of λ and that, for x and y belonging to K, if x is between the «neighbours» of y, y is between «neighbours» by x, and viceversa.

With reference to (1) there is little or nothing to add: the reflexivity of λ is intuitively clear and does not require further investigations. Instead, it should be appropriate to focus on the remaining constraints, because these are nontrivial aspects of locological spaces. I shall start with examining the last constraint then ending with the second. The order is not random but it is explained from the fact that a proper understanding of the «not thinness» of λ gives access to crucial properties of locological spaces that seem well adapted to intuitive (or «unorthodox») topology of notional domains.

7.5.2 *Non-trivial Symmetries …*

As for (3), I have not particular remarks: the reflexivity and symmetry of λ′ are acquired notions, I do not intend to proceed to a further examination of the last property assigned, i.e. namely the not thinness. It is, in fact, as for reflexivity, a property that λ′ shares with λ and of which I will discuss in reference to (2). The only point that I must discuss is relative to the fact that the «behaviour» of λ is different, in some respects, from the behaviour of λ′.

---

[21]De Glas and Desclés (1996, p. 30–31).

[22]Cf. De Glas and Plane (2005, p. 113): «Intuitivement, la relation λ peut être interprétée comme une relation de […] «proximité» […]».

For example: unlike λ', λ is not necessarily symmetrical—and it isn't, despite the fact that it «contains» λ'. Similar asymmetries are verified from some elementary locologies. The simplest structure provides as universe the real line, i.e. $X = \mathrm{R}$. If so, it can be imagined that, for at least two distinct points $x, y \in X$, λ $[x]$ and λ $[y]$ are closed intervals, respectively, with centre x and radius a_x (i.e. $\lambda[x] = [x - a_x, x + a_x]$) and with centre y and radius a_y (i.e. $\lambda[y] = [y - a_y, y + a_y]$)—where a_x, a_y are real positives and $a_x \neq a_y$.

For a choice of appropriate values, the verification of non-symmetry of λ is immediate. So: for $x = 3$ and $y = 4$ (with $a_x = 2$ e $a_y = 1/2$) it results that, in the closed interval with centre 3 and radius 2, 4 is among the neighbours of 3, i.e. $4 \in \lambda$ (3), while, in the closed interval with centre 4 and radius 1/2, 3 is not among the neighbours of 4, i.e. $3\ \lambda \notin (4)$. Hence, in fact, the non-symmetry of λ.

As De Glas states (see De Glas and Desclés 1996): «Que λ ne soit pas nécessairement symétrique peut se justifier sur un exemple très simple. Supposons que, sur $\mathrm{X} = \mathrm{R}$, on souhaite avoir $\lambda(x) = [x - a_x, x + a_x]$, où a_x est un réel positif. Alors, à moins que a_x ne soit constant, λ n'est pas symétrique». We have to examine (2).

7.5.3 … and Levels of Detail

As for (2), this is the most interesting constraint. From an intuitive viewpoint, it ensures that each locological space provides the transition from a coarser level of detail to the finer one: «[…] la relation non maigre λ […] représente l'idée intuitive du changement de grain».[23] This possibility is ensured by the fact that λ operates as a kind of map that assigns to each point of the universe the set of its neighbours (or *halo*). In particular: a map with domain X and codomain $\wp$ (X) (i.e. the power set of X).

As De Glas and Plane (2005: 113) state: «[…] par un petit abus de langage, λ désigne non seulement une relation binaire sur Ω (i.e. sur X, n.d.r.), mais aussi désormais l'application de Ω sur l'ensemble $\wp(\Omega)$ des parties de Ω qui à tout objet x associe son *halo*». With a metaphor: λ is a «lens» that zooms in over points of small enough regions of X and that «focuses» the points immediately close (or contiguous) to each point of each region of X. With the same metaphor: the points in close proximity to any point of the universe are, unless than zooming on the given point, indistinguishable from it: «[…] un *halo* […] peut être vu comme l'ensemble des points indistinguables de ce point […]».[24]

So, any zooming involves a change in the levels of detail of the locological space. Ghadakpour (2002: 54) provides, in that regard, an example based on the general idea that, in a locological space, the extremes of intervals are in turn

[23]Ghadakpour (2002, p. 54).

[24]De Glas and Plane (2005, p. 109).

intervals: «L'idée consiste à considérer les bornes d'un intervalle comme un nouvel intervalle».

Hence the change in the level of detail, in the sense that, under the action of the «lens» (i.e. of λ), each extreme of interval is associated with the corresponding *halo* and therefore «ce qui pouvait apparaître comme ponctuel à l'échelle de l'intervalle entier apparaît maintenant comme ayant un intérieur [...]».

Paraphrasing Desclés (2002), we move from «places without thickness» (or *punctual boundaries*) to «places whith thickness» (or *boundaries as zones*). The locology has the formal tools to clarify relevant differences. In this sense, the notions of «inner boundary» and «outer boundary» (cf. Desclés 2009) play a significant role. Let us see how.

7.5.4 Inner Boundaries and the Outer Boundaries

The intuitive idea of boundaries as «zone», is captured on the formal level from the locological notion of «thick boundary» (or *frontière épaisse*). This concept is used in reference to a mathematical entity that is the union of a inner boundary and an outer boundary: «La quasi-topologie introduit des lieux abstraits avec des «frontières internes» et des «frontières externes» [...].»[25] Such entities are already outside of general topology.

Topological boundaries have neither exterior nor interior, but separate interior from exterior: «Les frontières séparent les intérieures des extérieures».[26] In short, these are the main differences. For a more in-depth knowledge, it is necessary to clarify what is meant exactly by «inner boundary» and «outer boundary». For this purpose, I will introduce the locological notions of *core* and *shadow*.

7.5.4.1 Core and Shadow

In the lexicon of locology, the words «core» and «shadow» define sets with certain properties. More precisely: subsets of some universe, in this case X, defined by operators that have as domain and codomain the same object, i.e. $\wp$ (X), and taking in turn the names of «core» operator, (labeled with h), and «shadow» operator (labeled with s). Thus: h: $\wp(X) \rightarrow \wp(X)$ and s: $\wp(X) \rightarrow \wp(X)$.

At this general level, the core and the shadow of a subset A of X all take the form of subsets of X assigned to A by h and s, respectively. The core of A will be denoted by $h(A)$; the shadow of A will be denoted by $s(A)$.

[25]Desclés (2011, p. 35, n. 6).
[26]Desclés (2012, p. 10).

As Breysse and De Glas state (2007: 227): «[…] one defines the following two operators *h* and *s* which associate to any A in $\wp$ (X) its core and its shadow, respectively. More precisely, let *h*: $\wp$ (X) → $\wp$ (X) be the operator which associate to any A its core […]».

In more detail (and with a slightly different terminology): for any subset A of X, the core of A consists of all elements of X whose *halos* are included in A, thus: $h(A) = \{x \in X: \lambda[x] \subseteq A\}$; on the other hand, the shadow of A consists of all elements of X where each *halos* has at least one element in common with A, thus: $s(A) = \{x \in X: \lambda[x] \cap A\}$.

The inner boundary and the outer boundary of a given set, in this case A, are defined by means of the locological concepts of core and shadow. In particular: the inner boundary of A, labeled with ∂_{int} (A), is defined as the complement of core of A with respect to A, i.e. $\partial_{in}t(A) = A \backslash h(A)$; the outer boundary of A, labeled with $\partial_{ext}(A)$, is defined as the complement of A with respect to the shadow of A, i.e. $\partial_{ext}(A) = s(A) \backslash A$. So, the global boundary is the union of $\partial_{int}(A)$ and $\partial_{ext}(A)$. Clarified these formal details, it is necessary to specify the impact they have on metalinguistic repertory of TOPE.

7.6 Locology: Metalinguistic Implications

From a certain point of view, the impact is minimal: the works of TOPE referring implicitly or explicitly to the concept of thick boundary neglect (and sometimes ignore) the formal background from which it originates. However, it seems that the locological main categories (or at least some of them) have been transposed correctly. Among other things, there are contributions that excel in technical expertise and who have direct knowledge of theoretical foundational instances of locology. For example: essays by Jean-Pierre Desclés, that represent one of the most interesting modern ramifications of TOPE.

In addition, authors less technically involved have produced about more than respectable results. So, in the case of examined metalinguistic construction (the notional domain) it seems that a certain literature of reference reads the intuitive topology of notional domain in locological (or quasi-topological) terms. The main results of this reading are to be appreciated on both conceptual and topographic level (i.e. *diagrammatic representations*). Let us start with examining the first aspect.

7.6.1 Thick Boundary

The use of the concept of «thick boundary» has implied, at least between those scholars who have paid particular attention to the impact of locology on the metalinguistic level, the use of complementary concepts of inner and outer boundary.

Therefore, the articulation of the notional domains in interior, exterior and the boundary has been replaced by a new partition that provides, in addition to an interior and an exterior, a (thick) boundary divided into inner boundary and outer boundary:

«Sur le plan théorique, cette analyse [...] nous oblige à adopter, pour schématiser le domaine notionnel, une quasi-topologie de préférence à une topologie classique puisqu'elle suggère l'existence d'une frontière interne et d'une frontière externe dont l'union forme une frontière épaisse».[27]

The use of such notions in the theory of notional domains has, however had, as a consequence, the relative impairment of locological definitions and a substantial impoverishment of the conceptual repertoire on which they were made. In fact, in the examined works, there is a lack of any reference to the concepts of core and shadow, while inner boundary and outer boundary are defined in the metaphorical terms of zones which are adjacent (respectively) to the interior and exterior of the domain. This, for example, is what you can read in reference to interior boundary: «[...] la frontière interne du domaine, c'est à dire dans cette zone du domaine qui ne relève plus de l'interieur [...], mais qui est lui adjacente, sans relever non plus de l'exterieur».[28] However, the metaphorical style of TOPE remains, roughly, compliant to the theoretical system of locology; indeed, in this case, it allows specifying (intuitively) concepts that would otherwise be accessible only to restricted groups of scholars. The diagrams used as support to clarify these concepts play, in this respect, an important role.

7.6.2 *Diagrams*

The diagrams used by locological (or quasi-topological) readings of notional domains are different from diagrams introduced in Sect. 7.2.1.2 for some superficial characteristics that exemplify the difference of TOPE in relation to the use of certain basic concepts of general topology. In short (and in terms of 3.1): they represent the boundaries of notional domains as interstitial spaces (or *transition zones*) between interior and exterior. This way of representation captures at intuitive level the concept of thick boundary and allows to grasp the complementary concepts of inner boundary and outer boundary. Consider the case of two diagrams used by Desclés (2009: 37). The Fig. 7.1 shows how the locological types of notional boundary differ from the corresponding topological types: what separates interior from exterior is no longer an infinitely thin edge, but a zone (or interval) with a certain extension. The Fig. 7.2 shows the presence of other two «sub-zones»: the orange sub-zone which is adjacent to the interior (i.e. the inner boundary) and ochre one which is adjacent to the exterior (i.e. the outer boundary).

[27]Mellet and Monte (2009, p. 44).

[28]Ibidem p. 42.

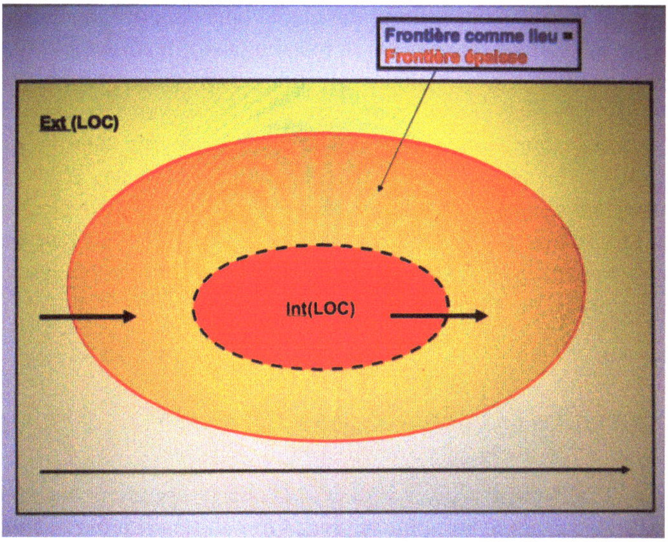

Fig. 7.1 The locological types of notional boundary differ from the corresponding topological types: what separates interior from exterior is no longer an infinitely thin edge, but a zone (or interval) with a certain extension

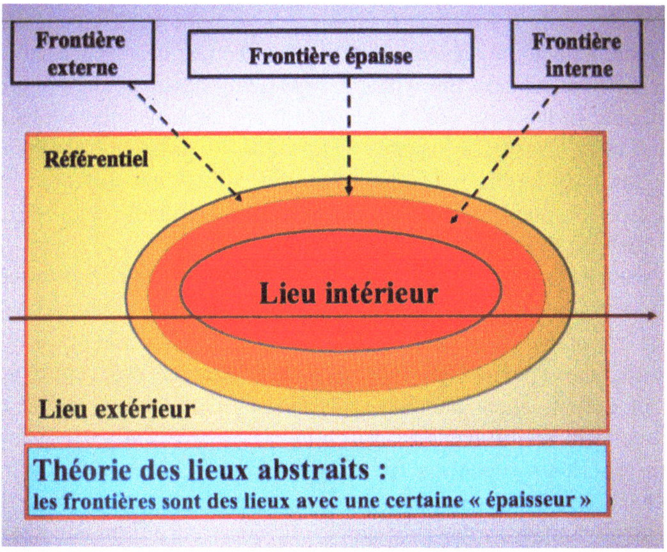

Fig. 7.2 The presence of other two «sub-zones»: the orange sub-zone which is adjacent to the interior (i.e. the inner boundary) and one ochre adjacent to the exterior (i.e. the outer boundary)

The union of the two zones shows the global boundary of the notional domain. From here, a qualitative refinement of metalinguistic representations associated with the classes of abstract occurrences of the notion: those occurrences with the highest number of properties in accordance with the organizing centre instantiate points of the inner boundary; others, instead, points of the outer boundary. The benefits of such diagrams are so important: they increase the level of detail of metalinguistic representations used in the descriptions of fragmented notion and they allow to visualize the locological concepts used within these descriptions. It only remains to explain how the intuitive topology of such representations would describe the situated occurrences of a notion.

7.7 Enunciative Occurrences As «Locators»

According to TOPE the text gives access to language operations that generate the classes of abstract occurrences of the notions. From the enunciative occurrences it is possible to trace back the mechanisms of identification and rejection of properties, which the genesis and recognition of possible exemplars of the notion are based on.

For example, an utterance like «That's a *really* fox!» (1) is the marker of a complex chain of linguistic operations that induces the construction of an abstract occurrence and its identification with the organizing centre of /fox/.

Moreover: from utterances like «That's *not* a fox!» (2), «That's *hardly* a fox» (3) or «It's *only* a fox» (4) it is possible to trace back chains of linguistic operations that induce the construction of other abstract occurrences and lay down for each a different relationship with the organizing centre of the notion.

In particular: in the case of (2), the underlying language operations induce the construction of an abstract occurrence that is radically different from the organizing centre of /fox/. As to the linguistic operations of (3) and (4), they induce the construction of abstract occurrences that are not identifiable with the organizing centre of /fox/ but not entirely different from it.

With some differences: in one case, the occurrence is presented as something that the speaker tends to assimilate to what is different from organizing centre—or from any generic exemplar of the notion. In the other, instead, occurrence is presented as something that the speaker assimilates, if not to the organizing centre, to a generic exemplar of notion. Some paraphrase of «hardly» and «only» in the considered utterances confirm the differences just outlined. From «That's hardly a fox!» you can switch to «That's vaguely a fox!», while from «That is only a fox» you can switch «That's a fox, nothing more».

Hence the assimilations that distinguish one occurrence from the other. The four utterances generate specific relationships that have a direct encoding on the notional domain of /fox/.

In brief: occurrences built from the first and second utterance instantiate, respectively, the central point within the domain and an external point. The occurrences built from the third and fourth, instead, instantiate, respectively, a point of the outer boundary and a point of the inner boundary. As Mellet and Monte affirm (2009: 42–46): «L'adverbe *juste* [...] positionne [...] l'occurrence à la frontière interne du domaine [...] *à peine* situe l'occurrence à l'intersection de la frontière avec l'extérieur du domaine [...]».

The intuitive topology (or quasi-topology) associated with a class of abstract occurrences of the notion allows to describe each of these utterances as locators—i.e. as linguistic constructions that show (on the textual level) a relationship between an occurrence of the notion and the point of an abstract space. In general terms, the enunciative occurrences of /fox/ (as the enunciative occurrences of any other notion) are the textual manifestation of *cognitive artefacts* that instantiate the points of an abstract space associated with the notion. In metaphorical terms, every enunciative occurrence gives instructions for computing a position in a notional space. To quote again Gerner (1995: 39–40): «Nous pouvons donc poser comme équivalents les termes occurrences et calcul de position».

7.8 Conclusions

The analysis conducted on one of the metalinguistic constructions of TOPE, the notional domain, allowed to examine three pairs of basic concepts: (a) *markers* and *notions*; (b) *fragmentation* and *construction of existence*; (c) *abstract occurrences* and *located occurrences*. In particular, it has helped to deepen the concepts of abstract occurrence and located occurrence offering two metalinguistic representations: (a) abstract occurrences as *points of abstract notional spaces*; (b) occurrences located as *position markers* on these spaces. I would close these notes with two summary observations on the general meaning of such representations.

7.8.1 The First Observation

The first observation is that they are formal tools that codify in terms of elementary intuitive topologies the language operations underlying the utterances' production. As for the fragmentation, this encoding offers a description of the *notional types* (or *organizing centres)* around which it is ordered the corresponding abstract occurrences. As for the construction of existence, it offers a description of the enunciative form of notional types and relationships between types and abstract occurrences. Of course, the two descriptions are two*sides of the same coin*. From the perspective of TOPE, the construction of an abstract occurrence is an integral part of the cognitive mechanisms of identification and rejection of properties on which is based the recognition of such occurrences. Hence, the relation between the two levels of representation: abstract occurrences as points of notional spaces and located occurrences as position markers on these spaces.

7.8.2 The Second Observation

The second observation is that they offer a description of the relationships between enunciative occurrences of a notion. For example: starting from a family of utterances like «That's a fox», «No, it's not at all a fox», «It's hardly a fox» etc., the intuitive topology associated with a class of abstract occurrences of /fox/ assigns to each utterance a position into the corresponding notional space. Moreover, from the analysis of these locations it is possible to ascend toward an enunciative universe—i.e. a concatenation of language operations that builds the existence of certain abstract occurrences and establishes, for each of them, a specific relationship with the organizing centre of the notion.

Nevertheless, I could not add more than this—but only to admit that the samples examined are greatly simplified reductions of the enunciative material on which TOPE normally works. What shall we say about modal modulations as «I guess it's a fox», «I think it's a fox» or figurative effects as «That woman is a fox», «He (or she) is intelligent as a fox», etc.? Of course, a lot could be said, and recent works on these issues (Ducard 2006) have been committed in the encoding analysis of such utterances. Nevertheless, this goes beyond the limits of this paper, and then I will not discuss—at least not in this article.

References

Bouscaren, J. et al. (1987). *Grammaire et textes anglais. Guide pur l'analyse linguistique*. Paris: Ophrys.

Breysse, O., & De Glas, M. (2007). A New Approach to the Concepts of Boundary and Contatcs: Toward an Alternative to Mereotopology. *Fundamenta Informaticae, 78*, 217–238.

Culioli, A. (1978a). The concept of notional domain, *Language Universals. Paper from the conference held at Gummersbach/Cologne, Germany, October 3–8, 1976. Tübingen: Narr*. In A. Culioli (2000) (pp. 67–81).

Culioli, A. (1978b). Valeurs modales et opérations énonciatives, *Le Français Moderne 46*(4), 135–155. Paris, In Culioli, (2000)

Culioli, A. (1981). Sur le concept de notion, *Bulag*. In A. Culioli, (2000) (pp. 47–65).

Culioli, A. (1986). La frontière, *Lignes de partage, Cahiers Charles V (8)*, In A. Culioli, (2000) (pp. 83–90).

Culioli, A. (1989). Representation, Referential processes, and Regulation. Language Activity as form production and recognition, In Culioli, A. (2000a): pp. 177–213.

Culioli, A. (1991). Structuration d'une notion et typologie lexicale. À propos de la distinction dense, discret, compact, *Bulag*. In Culioli A., (2000c) (pp. 9–15).

Culioli, A. (2000a). *Pour une linguistique de l'énonciation. Opérations et représentations, TOME I*. Paris: Ophrys.

Culioli, A. (2000b). *Pour une linguistique de l'énonciation, Formalisation et opérations de repérage, TOME II*. Paris: Ophrys.

Culioli, A. (2000c). *Pour une linguistique de l'énonciation, Domaine Notionnel, TOME III*. Paris: Ophrys.

Culioli, A. (2002). *Variations sur la linguistique*. Klinckisieck, Paris: Entretiens avec Frédérik Fau.

Culioli, A. (2014). *L'arco e la freccia. Scritti scelti di Antoine Culioli.* Con il saggio di Francesco La Mantia *Sul lessico della linguistica di Culioli.* Prefazione di Tullio De Mauro. Bologna: Il Mulino.

De Glas, M. (2009b) (éd.). *Le continu mathématique. Nouvelles conceptions, nouveaux enjeux,* Intellectica. Revue de l'Association pour la Recherche Cognitive, n. 51, Paris.

De Glas, M. & Desclés, J. P. (1996). Du temps linguistique comme idéalisation d'un temps phénoménal, *Intellectica*, 2(23), 159–192.

De Glas, M., & Plane, J. (2005). *Une approche formelle de la typicité.* Paris: CAHIERS DU CREA.

Desclés, J. P. (1997). Schémes, notions, prédicats et termes, *Mélanges offerts à Jean-Blaise Grize*, Louvaine.

Desclés, J. P. (1998). Opérations de qualification et quantification (pp. 1–9), In http://lalic.paris-sorbonne.fr/PAGESPERSO/Descles/.

Desclés, J. P. (2002). Quelques concepts empruntés par A. Culioli à la logique et aux mathématiques. In A. Culioli, (2002) (pp. 243–253).

Desclés, J. P. (2009). Schèmes sémantico-cognitifs et schèmes quasi-topologiques (pp. 1–149), In http://lalic.paris-sorbonne.fr/PAGESPERSO/Descles/.

Desclés, J. P. (2011). Mathématiser les concepts dans les sciences humaines, AEIS (pp. 1–36), In http://www.science-inter.com/Emergence/TMSSDesclesTexte.doc.

Desclés, J. P. (2012). Du Trimorphe aux frontières quasi-topologiques, In M. Becquelin, (2012) (pp. 38–68).

Ducard, D. (2006). *Entre grammaire et sens.* Paris: Ophrys.

Ducard, D. (2009). Le graphe du geste mental dans la théorie des opérations énonciatives d'A. Culioli , Cahiers Parisiens, *Parisian Notebooks 5*, 555–576, The University of Chicago Center in Paris.

Dufour, F. (2011). Du spectacle linguistique aux opérations cognitivo-discursives, *Revue interdisciplinaire Textes et contextes [en-ligne] 5*, http://revuesshs.u-bourgogne.fr/textesandcontextes/document.

Filippi-Deswelle, C. (2008). Le domaine notionnel d'Antoine Culioli (ou comment la géo.graphie (méta)linguistique opérante dans nos représentations topologiques abstraites régule notre activité langagière). In L. Villard (éd,) (pp. 299–314).

Ghadakpour, L. (2002). *Le système conceptuel, à l'interface entre le langage, le raisonnement, et l'espace qualitatif :vers un modèle de représentations éphémères*, Thèse présentée pour obtenir le diplôme de Docteur de l'École Polytechnique, Paris.

Gerner, M. (1995). Pour une typologie discursive via la théorie des esquisses. *Mathématiques et sciences humaines, 129*, 37–69.

Janich, K. (1994). *Topologia.* Bologna: Zanichelli.

Lipschutz, S. (1969). *General Topology.* New-York: McGraw-Hill.

Mellet, S. (2009a). La frontière notionnelle en langue et en discours. In Mellet S. (éd.) (2009) (pp. 7–20).

Mellet, S. (2009b) (éd.). *La frontière notionnelle en langue et en discours, Cahiers de Praxématique, 53*, Paris.

Mellet, S., & Monte, M. (2009). Juste/à peine et la construction de la frontière notionnelle. In Mellet S. (éd.) (2009) (pp. 33–55).

Vignaux, G. (1988). *Le discours Acteur du Monde.* Paris: Ophrys.

Part III
Linguistic Complexity: Physics, Computation and Biology

Chapter 8
Fiat Lux Versus Fiat Lumen: Quantum Effects in Linguistic Operations

Fortunato Tito Arecchi

8.1 A Methodological Premise

Classifying the information content of neural spike trains, an uncertainty relation emerges between the bit size of a word and its duration. This uncertainty is ruled by a quantum constant whose numerical value has nothing to do with Planck's constant. A quantum conjecture explains the onset and decay of the memory window connecting successive pieces of a linguistic text. The conjecture here formulated is applicable to other reported evidences of quantum effects in human cognitive processes, so far lacking a plausible framework since no efforts to assign a quantum constant have been associated.

Any scientific description entails a complementarity between its Extension and Detail. The two aspects of a scientific description, namely, Extension and Detail, result mutually conflicting.

For instance, a successful approach explaining the cosmic evolution from the Microwave Background to the Galaxy formation cannot explain the details of planet differentiation, why e.g., the Earth has a magnetic field providing the Van Allen belt shield from solar particles, or the water necessary for life. Two different sets of foundational principles must be introduced in order to explain the two classes of phenomena, that is, the fundamental objects of Planetology and their mutual interactions must be introduced appropriately and cannot be derived from the general principles of Cosmology.

In a similar way, a general powerful QFT (quantum field theory) approach has been developed to explain the brain and memory organization starting from the collective organization of water dipole quanta in living matter (Vitiello 2001; Freeman and Vitiello 2006).

F.T. Arecchi (✉)
University of Florence, Florence, Italy
e-mail: tito.arecchi@ino.it

F. La Mantia et al. (eds.), *Language in Complexity*,
Lecture Notes in Morphogenesis, DOI 10.1007/978-3-319-29483-4_8

Even though, the puzzling fact that in a human linguistic endeavor, words are mutually influencing through their meanings, so that an "infinite use emerges from a finite amount of resources" (Humboldt 1836) has no explanation whatsoever in the QFT of brain and memory, that has no tools to differentiate the human brain behavior from that of other animals.

In this work, we present a specific theoretical approach that provides that sound explanation for the linguistic performances is not achievable in the QFT of brain and memory.

8.2 Two Separate Cognitive Processes

In (Arecchi 2012), I have analyzed two distinct moments of human cognition, namely, apprehension (A) whereby a coherent perception emerges from the recruitment of neuronal groups, and judgment (B) whereby memory recalls previous (A) units coded in a suitable language, these units are compared and from comparison it follows the formulation of a judgment.

The first moment (A) has a duration around 1 s; its associated neuronal correlate consists of the synchronization of the EEG (electroencephalographic) signals in the so-called gamma band (frequencies between 40 and 60 Hz) coming from distant cortical areas. It can be described as an interpretation of the sensorial stimuli on the basis of available algorithms, through a Bayes inference.

Precisely (Arecchi 2012), calling h (h = hypothesis) the interpretative hypotheses in presence of a sensorial stimulus d (d = datum), the Bayes inference selects the most plausible hypothesis h^*, that determines the motor reaction, exploiting a memorized algorithm $P(d|h)$, that represents the conditional probability that a datum d be the consequence of an hypothesis h.

The $P(d|h)$ have been learned during our past; they represent the equipment whereby a cognitive agent faces the world. By equipping a robot with a convenient set of $P(d|h)$, we expect a sensible behavior. The second moment (B) entails a comparison between two apprehensions (A) acquired at different times, coded in a given language and recalled by the memory. If, in analogy with (A), we call d the code of the second apprehension and h^* the code of the first one, now—at variance with (A)—h^* is already given; instead, the relation $P(d|h)$ which connects them must be retrieved, it represents the conformity between d and h^*, that is, the best interpretation of d in the light of h^*.

Thus, in linguistic operations, we compare two successive pieces of the text and extract the conformity of the second one on the basis of the first one. This is very different from (A), where there is no problem of conformity but of plausibility of h^* in view of a motor reaction.

Let us make two examples: a rabbit perceives a rustle behind a hedge and it runs away, without investigating whether it was a fox or just a blow of wind. On the contrary, to catch the meaning of the 4th verse of a poem, we must recover the meaning of the 3-d verse of that same poem, since we do not have a priori algorithms to provide a satisfactory answer.

Once the judgment, that is, the $P(d|h)$ binding the codes of the two linguistic pieces in the best way, has been built, it becomes a memorized resource to which to recur whenever that text is presented again. It has acquired the status of the pre-learned algorithms that rule (A).

However—at variance with mechanized resources—whenever we reread the same poem, we can grasp new meanings that enrich the previous judgment $P(d|h)$. As in any exposure to a text (literary, musical, figurative) a rereading improves our understanding.

(B) requires about 3 s and entails self-consciousness, as the agent who expresses the judgment must be aware that the two successive apprehensions are both under his/her scrutiny and it is up to him/her to extract the mutual relation.

As a fact, exploring human subjects with sequences of simple words, we find evidence of a limited time window around 3 s (Pöppel 1997, 2004), corresponding to the memory retrieval of a linguistic item in order to match it with the next one in a text flow (be it literary, or musical, or figurative).

At variance with (A), (B) does not presuppose an algorithm, but rather it builds a new one through an inverse Bayes procedure (Arecchi 2007). This construction of a new algorithm is a sign of creativity and decisional freedom. Here the question emerges: can we provide a computing machine with the (B) capacity, so that it can emulate a human cognitive agent? Turing (1950). The answer is NOT, because (B) entails non-algorithmic jumps, insofar as the inverse Bayes procedure generates an ad hoc algorithm, by no means preexistent.

Figures 8.1, 8.2, 8.3, and 8.4 that follow and their captions explore in detail these aspects (Pöppel 1997).

After having shown evidence of this short-term memory window bridging successive pieces of a linguistic text, we formulate a quantum conjecture. This conjecture fulfills two needs, namely, (i) explaining the fast search in a semantic space, whose sequential exploration by classical mechanisms would require extremely long times, incompatible with the cadence of a linguistic presentation

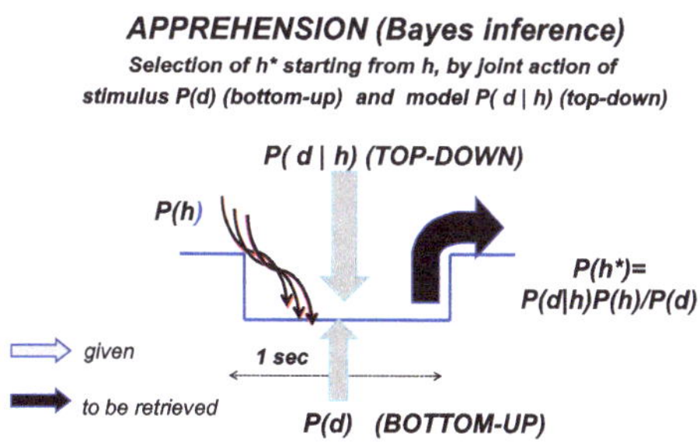

Fig. 8.1 Apprehension as a Bayes inference. One formulates a manifold of hypotheses; each one provides a datum through the *top-down* conditional probability; only the hypothesis that generates the actual datum (*bottom-up*) is the plausible one

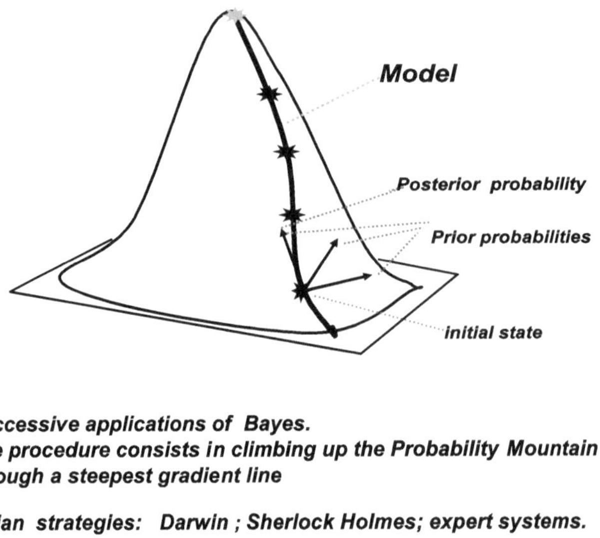

Fig. 8.2 Recursive application of Bayes is equivalent to climbing a probability mountain, guided by the model, that is, the conditional probability that a hypothesis generates a datum. This strategy is common e.g., to Darwin evolution and to Sherlock Holmes criminal investigation; since the algorithm is unique, it can be automatized in a computer program (expert system)

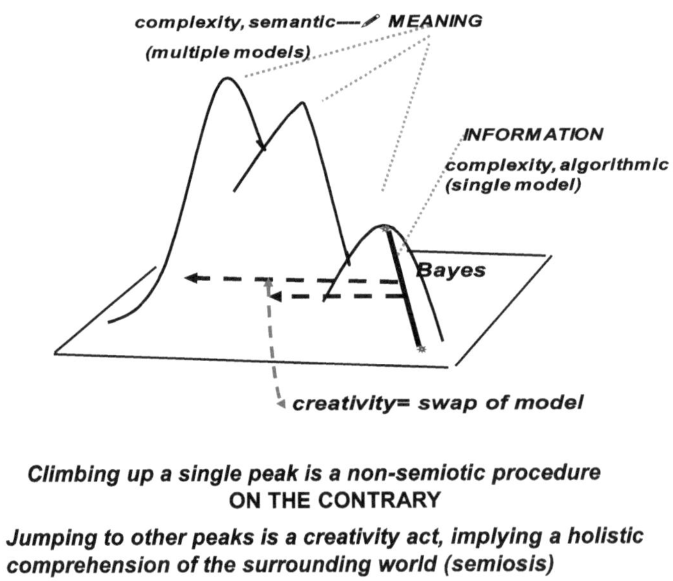

Fig. 8.3 Comparison of two different complexities, namely, (i) the algorithmic c., corresponding to the bit length of the program that enables the expert system to a recursive Bayes; and (ii) semantic c., corresponding to the occurrence of different models (provided they are countable); in fact they are not, because we will see that different meanings result from a quantum exploration

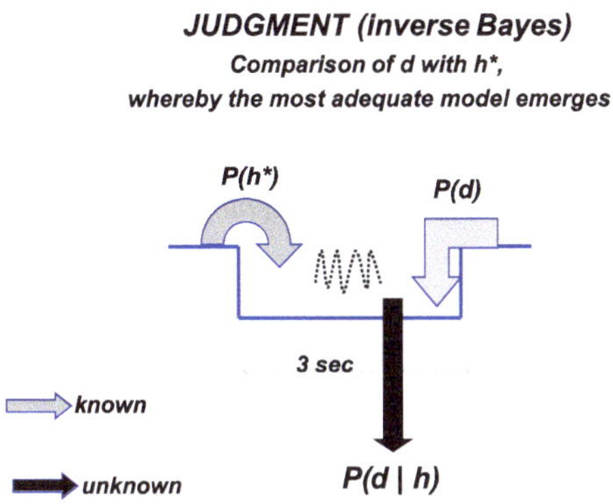

Fig. 8.4 The inverse Bayes procedure that occurs in linguistic endeavors, whereby a previous piece of a text is retrieved by the short-term memory and compared with the next one: the appropriate conditional is no longer stored permanently but it emerges as a result of the comparison (judgment and consequent decision)

(Grover 2001); (ii) introducing a fundamental uncertainty ruled by a quantum constant that yields a decoherence time fitting the short-term memory window.

The memory enhancement associated with linguistic flows is an exclusively human operation, not applicable to a cognitive agent that operates recursively, exploiting algorithms already stored in the memory. If the conjecture will be confirmed, the quantum mechanism would explain the a posteriori construction of novel interpretational tools.

Classifying the information content of spike trains, an uncertainty relation emerges between the bit size of a word and its duration. This uncertainty is ruled by a quantum constant that can be given a numerical value and that has nothing to do with Planck's constant. A quantum conjecture might explain the onset and decay of the memory window connecting successive pieces of a linguistic text. The conjecture here formulated is applicable to other reported evidences of quantum effects in human cognitive processes, so far lacking a plausible framework since no efforts to assign a quantum constant have been associated.

Models of quantum behavior in language and decision taking have already been considered by several authors but without a dynamical basis, starting 1995 (Aerts and Aerts 1995; Aerts 2009); and over the past decade (Khrennikov 2010). Most references are collected in a recent book (Busemeyer and Bruza 2012). None of these authors worries about the quantum constant that must replace Planck's constant.

However, a quantum behavior entails pairs of incompatible variables, whose measurement uncertainties are bound by a quantization constant, as Planck's in the original formulation of Heisenberg. One cannot apply a quantum formalism without having specified the quantum constant ruling the formalism. For this reason, all reported quantum attempts must be considered flawed, because (i) either they overlook the need for a quantization constant (Aerts and Aerts 1995; Aerts 2009; Khrennikov 2010), or (ii) use Planck constant and consequently arrive to very short decoherence times, incompatible with cognitive processes (Busemeyer and Bruza 2012; Penrose 1994; Hagan et al. 2002; Tegmark 2000).

After stressing in Sect. 8.3 the difference between (A)-perception and (B)-linguistic processes, we devote Sect. 8.4 to the quantum aspects of an interrupted spike train, that provide a non-Newtonian quantization suitable for the foundation of quantum linguistic processes.

8.3 Perceptions Versus Linguistic Processes

We have distinguished between two different cognitive processes, namely:

(A) Perception, whereby a sensorial stimulus is interpreted in terms of "models," or behavioral patterns, already stored in the long-term memory; the interpreted stimulus elicits a motor reaction; duration from a few hundred milliseconds up to 1 s; adequately described as a Bayesian procedure; common to all animals, and

(B) Linguistic processes, only human, whereby a sequence of pieces, coded as words of the same language, are sequentially presented to the cognitive agent; each piece is interpreted in terms of the previous one recovered by the short-term memory; such a comparison must be performed within 3 s; otherwise, the sequence must be repeated.

Focusing on (b), a decision, or judgment, is the interpretation of the last piece based upon the meanings of the previous one. Scanning all possible meanings of each piece entails a fast search process that requires a quantum search.

Plenty of approaches have tackled quantum-like aspects of language processing (Aerts and Aerts 1995; Aerts 2009; Khrennikov 2010; Busemeyer and Bruza 2012); however these approaches either did not discuss limitations due to a quantum constant, hence, they are purely formal without a physical basis or they refer to the quantum behavior of Newtonian particles (Penrose 1994; Hagan et al. 2002) and hence are limited by a coherence time estimated around 10–14 s. (Tegmark 2000; Koch and Hepp 2006) well below the infra-sec scale of the cognitive processes (Rodriguez et al. 1999).

8.4 A Novel Aspect of Quantum Behavior

Standard quantum physics emerges from the Newtonian physics of a single particle. Refer for simplicity to 1-dimension. The uncertainties of position x and momentum p obey the Heisenberg condition

$$\Delta x \Delta p \geq \hbar . \quad (8.1)$$

All quantum formalism is a consequence. For instance, comparison with the Fourier condition

$$\Delta x \Delta k \geq 1 \quad (8.2)$$

suggests the DeBroglie relation

$$k = p/\hbar \quad (8.3)$$

whence the single particle interference, which contains the only quantum mystery (Feynman 1966) and Schroedinger wave equation.

In the x, p space, instead of Euclidean points $\Delta x = 0$, $\Delta p = 0$, we have uncertainty rectangles; thus the uncertainty areas of two separate particles can overlap: this is the origin of entanglement.

Now, let us consider a non-Newtonian phenomenon consisting of a temporal train of identical spikes of unit area and duration τ_b = 3 ms (bin) positioned at unequal times. This is a sound model for the electrical activity of a cortical neuron (Rieke et al. 1996; Arecchi 2004a, b). The corresponding signal is a binary sequence of 0's and 1's, depending on whether a given bin is empty or filled by a spike. Spike synchronization, i.e., temporal coincidence of 0's and 1's, is considered as the way cortical neurons organize in a collective state (Singer and Gray 1995) (Fig. 8.5).

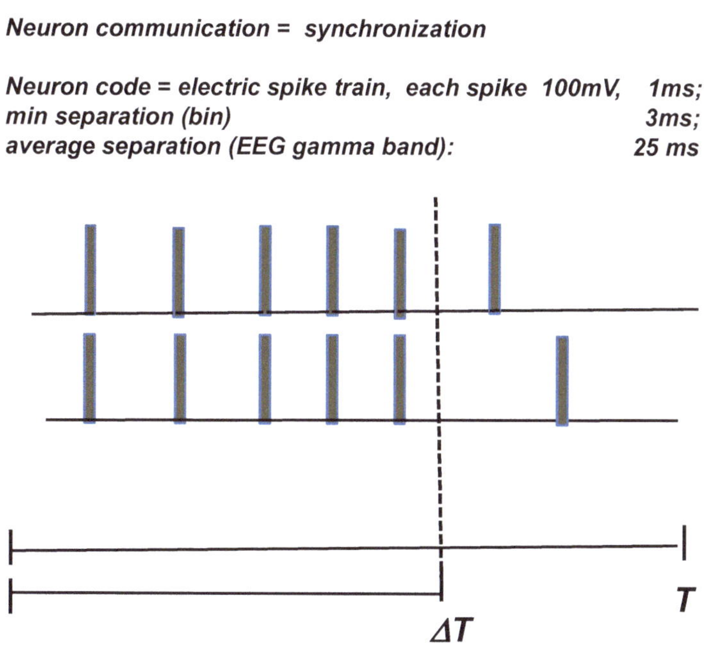

Fig. 8.5 Two spike trains of duration T, synchronized up to ΔT. The number of different realizations is $2^{(T-\Delta T)}$

Each cortical neuron has two ways to modify its spike occurrence, namely, either coupling to other cortical neurons or receiving signals from extra-cortical regions.

Let us take a processing time $T = 300$ ms, then, the total number of binary words that can be processed is $P_M = 2^{300/3} \cong 10^{33}$. At the end of a computational task, a decision center [called GWS = global workspace (Baars 1988; Dehaene et al. 2003)] picks up the information of the largest synchronized group and—based upon it—elicits a decision.

In the *perceptual* case (A), the cognitive action combines a bottom-up signal provided by the sensorial organs with a top-down interpretation provided by long-term memories stored in extra-cortical areas (Grossberg 1995).

In the *linguistic* case (B), the comparison occurs between the code of the second piece and the code of the previous one retrieved by the short-term memory. Here, we should consider a fact which so far had escaped a full explanation. Namely, spikes occur at average rates corresponding to the so-called EEG gamma band (say, around 50 Hz, that is, average separation 20 ms) (Rieke et al. 1996). However, superposed to the gamma band, there is a low-frequency background (theta band, around 7 Hz), which controls the number of gamma band bursts (Jensen and Colgin 2007). We show that interruption of a spike train introduces a quantum uncertainty, hence an entanglement among different words. This entanglement provides a fast quantum search of meanings that in classical terms would take a much longer time.

The theta-gamma cross-modulation corresponds to stopping the neural sequence at $\Delta T \leq T$. As a result, all spike trains equal up to ΔT, but different by at least one spike in the interval $T - \Delta T$, provide an uncertainty cloud ΔP such that (Arecchi 2003, 2004a, b):

$$\Delta P = 2^{(T-\Delta T)/\tau} = P_M 2^{-\Delta T/\tau} \tag{8.3}$$

Thus we have a peculiar uncertainty of exponential type between spike information P and duration T, that is,

$$\Delta P \cdot 2^{\Delta T/\tau} = P_M \tag{8.4}$$

By a variable change

$$y = 2^T \tag{8.5}$$

we arrive to a product-type uncertainty relation

$$\Delta P \Delta y = P_M \tag{8.6}$$

In the space (P, y) we have a Heisenberg-like uncertainty relation. Following the standard procedure of a quantum approach, we expect single-particle interference and two-particle entanglement in such a space

For $\Delta P = 1$ (minimal disturbance represented by 1 spike) we have the *decoherence* pseudo time $\Delta y_d = P_M$.

The corresponding decoherence time (in bins) is

$$\text{decoherence} - \text{time} = \log_2 P_M \equiv 100 \cdot (\text{bins}) \tag{8.7}$$

and going from bins to sec:

$$\text{decoherence time} = 0.3\,\text{s} \tag{8.8}$$

very far from the naive value of 10^{-14} s evaluated for Newtonian particles disturbed by the thermal energy $k_B T$ at T = room temperature (Tegmark 2000).

8.5 Conclusions

Innovations brought about by the linguistic processes in the brain: (1) The quantum constant for spike number–duration uncertainty has nothing to do with Planck's constant, a new type of quantum behavior has to be considered; spike synchronization is a peculiar physical process that cannot be grasped in terms of Newtonian position-momentum; (2) The energy disturbance which rules the decoherence time is by no means $k_B T$ (k_B being Boltzmann constant and T the room temperature), but it is replaced by the minimal energy necessary to add or destroy a cortical spike. This energy corresponds to the opening along the axon of about 107 ionic channels each one requiring an ATP → ADP + P reaction involving 0.3 eV, thus the minimal energy disturbance in neural spike dynamics is around 108 $k_B T$ (Arecchi 2004a, b). This is the evolutionary advantage of a brain: to live comfortably at room temperature and be barely disturbed, as it were cooled at 10^{-8} the room temperature.

As for the interpretation (hermeneutics) of a cognitive experience (be it perceptual or linguistic), we represent in Fig. 8.6 the procedural interpretation by a computing machine (CIRCLE) against that of any human language (COIL).

As for the *CIRCLE*, in information science, an *ontology* is a formal definition of the properties, and interrelationships of the entities that exist for a particular domain of discourse. An ontology compartmentalizes the variables needed for some set of computations and establishes the relationships between them. For instance, the booklet of the replacement parts of a brand of car is the ontology of that car. The fields of artificial intelligence create ontologies to limit complexity and to organize information. The ontology can then be applied to problem solving. Nothing is left out; we call this cognitive approach "finitistic" as nothing is left out beyond the description; no new LUX is provided by repeated trials.

On the contrary, in any human linguistic endeavor (be it literary, or musical or figurative) A starts building a provisional interpretation $A1$ of the text; whenever A returns to B, he/she has already some interpretational elements to start with, and from there A progresses beyond, grasping new aspects $B2$, $B3$ … and hence going

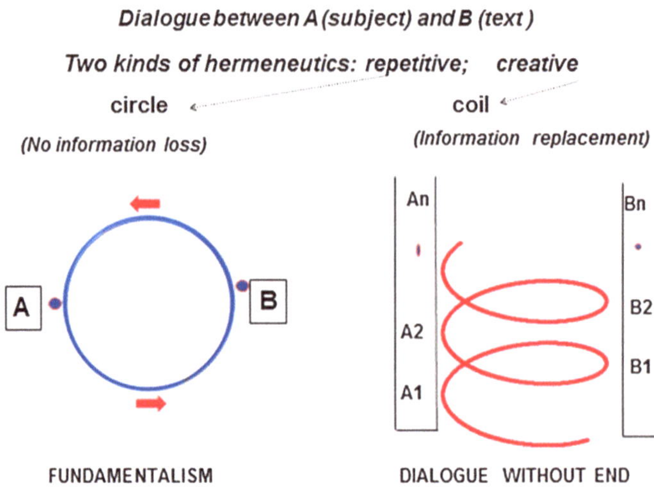

Fig. 8.6 Visual comparison between two kinds of interpretation of a text, or hermeneutics, namely, the CIRCLE, whereby the interpreter *A* attributes a finite and fixed set of meanings to the text *B*, and the COIL, whereby *A* captures some particular aspects of *B* and—based on that information—*A* approaches again the text *B* discovering new meanings. The novel insight provided at each coil is an indication of how LUX provides new semantic apertures

to *A*2 and so on (*COIL*). If *B* is not just a linguistic text, but another human subject, then *B* undergoes similar hermeneutic updates as *A*; this is a picture of the dialogical exchange between two human beings.

References

Aerts, D. (2009). Quantum structure in cognition. *Journal of Mathematical Psychology, 53*(5), 314–348.

Aerts, D., & Aerts, S. (1995). Applications of quantum statistics in psychological studies of decision processes. *Foundations of Science, 1*, 85–97.

Arecchi, F. T. (2003). Chaotic neuron dynamics, synchronization and feature binding: Quantum aspects TH. *Mind and Matter, 1*, 15–43.

Arecchi, F. T. (2004a). Chaotic neuron dynamics, synchronization and feature bindingc. *Physica, A, 338*, 218–237.

Arecchi, F. T. (2004b). *Caos e complessità nel viventeH*. Pavia: IUSS Press.

Arecchi, F. T. (2007). Complexity, information loss and model building: From neuro- to cognitive dynamics. In *SPIE Noise and Fluctuation in Biological, Biophysical, and Biomedical Systems* (pp. 6602–36).

Arecchi, F. T. (2012). Fenomenologia della coscienza: dall'apprensione al giudizio. in A. Ales Bello (Ed.) *...e la coscienza?Fenomenologia, psico-patologia, neuroscienze* (pp. 841–875). Laterza, Bari: Patrizia Manganaro.

Arecchi, F. T. (2013). La coscienza e il linguaggio: perché e come. Congetture quantistiche. In L. Urbani Ulivi (Ed.), *Strutture di mondo*, Vol. II, (pp. 199–226). Bologna: il Mulino.

Baars, B. J. (1988). *A cognitive theory of consciousness*. Cambridge: Cambridge University Press.

Busemeyer, J. R., & Bruza, P. D. (2012). *Quantum models of cognition and decision*. Cambridge: Cambridge University Press.

Dehaene, S., Sergent, C., & Changeux, J.-P. (2003). A neuronal network model linking subjective reports and objective physiological data during conscious perception. *Proceedings of the National Academy of Science (USA), 100*(14), 8520–8525.

Feynman, R. (1966), *The feynman lectures on physics,* Vol. III, Chapter 1.

Freeman, W., & Vitiello, G. (2006). Nonlinear brain dynamics as macroscopic manifestation of underlying many-body dynamics. *Physics of Life Reviews, 3*, 93–118.

Grossberg, S. (1995). The attentive brain. *American Scientist, 83*, 438–449.

Grover, L. K. (2001). From Schrödinger's equation to quantum search algorithm. *American Journal of Physics, 69*, 769–777.

Hagan S., Hameroff S. R., & Tuszynski J. A. (2002). Quantum computation in brain microtubules: Decoherence and biological feasibility. *Physical Review*, *65*(6).

Humboldt, W. (1836). *On language: On the diversity of human language construction and its influence on the mental development of the human species* (P. Heath, Trans.). Cambridge: Cambridge University Press, 1988.

Jensen, O., & Colgin, L. L. (2007). Cross-frequency coupling between neuronal oscillations. *Trends in Cognitive Science, 11*, 267–269.

Khrennikov, A. Y. (2010). *Ubiquitous quantum structure: From psychology to finance*. Berlin: Springer.

Koch, C., & Hepp, K. (2006). Quantum mechanics in the brain. *Nature, 440*, 611.

Penrose, R. (1994). *Shadows of the mind*. New York: Oxford University Press.

Pöppel, E. (1997). A hierarchical model of temporal perception. *Trends in Cognitive Sciences, 1*, 56–61.

Pöppel, E. (2004). Lost in time: A historical frame, elementary processing units and the 3-second window. *Acta Neurobiologiae Experimentalis, 64*, 295–301.

Rieke, F., Warland, D., de Ruyter van Steveninck, R., & Bialek, W. (1996). *Spikes: Exploring the neural code*. Cambridge: MIT Press.

Rodriguez, E., George, N., Lachaux, J. P., Martinerie, J., Renault, B., & Varela, F. (1999). Perception's shadow: Long-distance synchronization in the human brain. *Nature, 397*, 340–343.

Ronchi, V. (1957). *Optics. The science of vision*. New York: Dover Publications.

Singer, W., & Gray, C. M. (1995). Visual feature integration and the temporal correlation hypothesis. *Annual Reviews of Neuroscience, 18*, 555–586.

Tegmark, M. (2000). The importance of quantum decoherence in brain processes. *Physical Review*, E *61*, 4194–4206.

Turing, A. (1950). Computing machinery and intelligence. *Mind, 59*, 433–460.

Vitiello, G. (2001). *My double unveiled*. Amsterdam: John Benjamins.

Chapter 9
Two Ways into Complexity

Andrea Zeppi

9.1 Introduction

The dynamic hypothesis (DH) about cognition has often been presented as an alternative to the widely popular Computational Hypothesis (CH) in cognitive science. While the theoretical distance that separates these two approaches may seem to be significant, there are reasons, we argue, to reconsider the relationship between the dynamical and computational ways of understanding cognition. Here we will claim that CH and DH, rather than being competitors, are complementary framework in the explanation of cognition. This, we suggest, becomes evident when we consider that the two different notion of complexity that these hypothesis about cognition can be related to are very different indeed, but not contradictory. DH, we recognize, uses a notion of complexity derived from dynamical systems theory that seems to points toward psychological plausibility. On the other hand, CH may appeal to a computational notion of complexity that introduces elements of cognitive plausibility in the theoretical framework of cognitive science. In the following paragraphs, in order to test this last hypothesis, we will proceed as follows:

First, we will consider how the debate between dynamicists and computationalists, started in the first place and what are the main features of DH and of its notion of complexity;
Then we will see how CH evolved and consider how complexity can be recognized inside the CH framework;
Then we will see how computational complexity applies to cognition by linking computational tractability with cognitive plausibility;

A. Zeppi (✉)
Department of Cognitive Science, University of Messina, Via Concezione 8, 98121 Messina (ME), Italy
e-mail: azeppi@unime.it

F. La Mantia et al. (eds.), *Language in Complexity*,
Lecture Notes in Morphogenesis, DOI 10.1007/978-3-319-29483-4_9

In the end, we will argue that if we take the two different notions of complexity by comparison there are elements to consider them not adversary on the same ground, but as complementary model of explanation.

9.2 A Dynamical Cognition

The computational theory of mind, or computational hypothesis (CH from now on), has often been regarded as the received view in cognitive science. That comes from the fact that it provided, through its analogy between the working principles of the mind/brain and that of artificial computers, one of the very first tangible model of cognition. Instead of considering the riddle of criticisms[1] that it has received during history of cognitive science, we will concentrate on the debate between the supporters of CH and those philosophers and scientists that have proposed that a cognitive system should be intended as a dynamical system (DH).

The promoters of DH start from the premises that understanding natural cognitive system as dynamical models is radically different from considering them as computational systems. However, it should be clarified that DH supporters are not interested in criticizing any version of CH, but they often concentrate on that hypothesis that has been called the "paradigm of the computer" (PoC) (Cordeschi and Frixione 2007). The PoC could be defined as a pretty restrictive take on computationalism that can be roughly synthesized by following statement: a natural cognitive system works following the same principles of digital computers. This is clearly restrictive, since it presupposes that computationalism always intends cognitive systems as Von Neumanesque general-purpose architectures. However, what we actually want to do is evaluate DH by considering the features of the explanatory model that it proposes and not in relation to how it defines its polemical target. So, for now we will concentrate on those elements that introduced genuine novelty in the cognitivistic debate.

First it should be noted that DH is actually a twofold hypothesis (van Gelder 1998) about the nature of cognitive system. The first weaker sense in which DH can be interpreted is called the "nature hypothesis" and claims that cognitive agents actually instantiate dynamical systems in those parts that are considered to be responsible for cognitive performances. The second takes the name of "knowledge hypothesis" and claims that natural cognitive systems[2] should not only be considered as dynamical systems, but they should also be modeled after dynamical

[1]In this paper we will not address the following critical points: the semantics/syntax distinction (Putnam 1960; Searle 1980, 1992) or the computability tout-court of cognitive processes (Penrose 1999).

[2]Since artificial cognitive system may well be computational, DH is a theoretical hypothesis only over the nature of cognitive systems that are found in nature and that have therefore naturally evolved.

systems. These two interpretations of the term dynamicism both make use of the notion of dynamical system. The notion of dynamical system, as often happens in cognitive science, is a borrowed one and it comes from mathematics and science where it has already found numerous uses and definitions.[3] However, for what matters cognitive science, is often considered enough to say that dynamical systems are state-determined systems "with numerical states that evolve over time according to some rule" (van Gelder and Port 1998, p. 5). Where for "state-determined" is intended that the current state (the set of variables) happens inside a state pace[4] (or phase space) and that this state determines, given an evolution rule, the future states in which that system may evolve in time. This means that natural cognitive system and the phenomena that are ascribed to them should be intended and modeled utilizing the same concepts used for describing the growth of a population of bacteria, the behavior of an undamped pendulum or that of the solar system. There is no need for a sharp distinction between different processes, what is really important is the rate of change of the states of the system and the geometrical relations that develop in the state space. Everything cognitive has to be conceived, according to DH, as globally conceived and inherently relational. Systems are intended as a set of mutually bonded variables that evolve together in space (the environment) and time (they evolve simultaneously). These background features of the notion o dynamical systems have deep philosophical consequences and give to DH a set of very unique characteristics:

Emphasis on temporal evolution: dynamical systems' processes are not modeled around a notion of discrete state succession, but it emphasizes the rate of changes of the variable of the system in the unit of time. It is not the order to be important, but the changing aspects of the mutually linked variables. At the same time this approach enables to consider cognitive systems not as a diachronic architecture, but as system that instantiate processes always working simultaneously and in real time;
Emphasis on holism: with no distinction between central and peripheral processors, or between the brain and the mind, DH is a naturally embedded and embodied model of cognition (Thelen 1995). This means that, while there is no longer the need to sharply distinguish the mind from the brain and body, there is also no need to strongly separate cognitive systems from their environment. For these reason we can also consider DH as a naturally extended theory about cognition (Clark and Chalmers 1998). Furthermore a distributed, parallel, and nonmodular model of the mind can be easily conceived under DH, since there is no use for the notion of symbol and, therefore, for that of representation.

These same features are responsible of strongly binding DH with a notion of complexity. In the following paragraph, we will go through the reason why it is so and what are the main characteristics of such a notion.

---

[3]See van Gelder (1998, p. 618) for a series of examples of dynamical systems in physics and mathematics.

[4]The space where we find all the possible state in which the system may be in.

9.2.1 How About Complexity?

We have already seen how DH reimagines cognition as a continuous, interactive, and geometrically organized phenomenon. What is achieved this way is a theoretical hypothesis that naturally conceives cognition as a complex phenomenon and cognitive systems as inherently complex systems. DH does so, thanks to the relation between dynamic systems and dynamic system theory,[5] but also because under DH cognitive systems are considered as topological structures in which "the interaction among constituents of the system, and the interaction between the system and its environment, are of such a nature that the system as a whole cannot be fully understood simply by analyzing its components" (Byrne 1998, p. viii).

DH surely inherently presupposes a notion of complexity thanks to its emphasis on time evolution. The real-world decisions have to be taken in split seconds and actions, as grabbing a cup and taking a sip of water, and are needed to start and stop at just the right moment in order to succeed. All of that also seamlessly happens in complex continuum where a number of multiple processes all take place simultaneously and on different time-scale.[6] However, DH does not only pose an accent over the necessity of intending cognitive phenomena as real-time processes, but it introduces also the necessity of interpreting cognitive systems as real living beings acting in a living real world. The same actions and decisions of before are not taken by idealistic systems or architectures. Furthermore, rather than showing a computer-like architecture, they are organized in a net-like structure of interdependent elements which apparently does not behaves like any Von Neumann architecture we know about. As a matter of fact, neuromorphic models of information processing, such as the artificial neural networks of connectionists, are considered by dynamicists example of dynamical systems[7] (van Gelder and Port 1998; van Gelder 1995, 1998; Horgan and Tienson 2006). Understanding the CNS as a "single dynamical system with a vast number of state variables" (van Gelder and Port 1998, p. 34) comes pretty natural under DH.

Inside the dynamicist framework there has seen application in the modelisation of decision processes (Townsend and Busemeyer 1995) and sensorimotor activity (Saltzman 1995). Linguistics DH has made feasible a semiophysical (Petitot 1992; Thom 1988) approach to some of the trickiest feature of natural language, such as linguistic meaning, compositionality, semio-genesis, and lexical polysemy. Especially, this last field of enquiry reveals another characteristics of the dynamical

[5]A branch of pure mathematics concerned with the behavior of complex systems (Alligood et al. 1997).

[6]Since there is no discrete state subdivision different processes may well occur simultaneously but with different rates of change in the phase space.

[7]However, while dynamicists are keen to aknowledge artificial neural network as genuine dynamical systems, they often consider connectionism itself as an unfinished attempt at overcoming computationalism (van Gelder and Port 1998, p. 32).

framework: qualitative and phenomelogical aspects of cognition can be naturalistically explained (Petitot 1999) without the need for a reduction or an elimination. The flexibility of dynamical systems makes at least analogical explanations possible, so "even without an elaborate data time series, one can study a mathematical model which exhibits behavior that is at least qualitatively similar to the phenomena being studied" (van Gelder and Port 1998, p. 16).

In conclusion, we can now safely affirm that complexity is indeed a fundamental element of DH. Cognitive systems qualify as complex systems when considered dynamically because the model is closer to the reality of actual cognitive agents. By doing so DH places cognitive systems in real time and poses it in relation with a naturally complex environment. What complexity does in DH helping at closing the gap between the mind and body. Cognitive phenomena do not have a set of (computational) restraining them and the same goes for the actual architecture of the cognitive systems. Avoiding the constraints of computationalism helps at reconsidering the way in which a cognitive system is said to be cognitive in the first place. A dynamical cognitive system qualifies as cognitive thanks to what it does and not because of what it is. The dynamicist model does not define cognition in top-down way, but instead is the model that has to adapt in order to reach a better understanding of the explanandum. This gives the notion of complexity in dynamicism a quite specific flavor. In DH complexity is introduced not only in order to close the gap between models and reality, but also because there is a clear accent toward maximizing psychological plausibility. The actual constraints and limitations of the cognitive system are somewhat overlooked in order to reach a deeper understanding of those cognitive facts that were previously considered as irreducible or subject to plain elimination. While computationalism apparently focuses on finding constraints in order to keep the complexity of cognition at check, dynamicism exploits cognition in order to make our explanation of it closer to the actual experience of a mind.

In the following paragraph we will instead see how complexity works in the computationalist side of cognitive science. Contemporary computationalism complexity surely plays quite an important part but with very precise characteristics.

9.3 The Evolution of Computationalism

In the first paragraph, we have seen how criticisms coming from DH supporters focus on what we have called the PoC version of computationalism. We also said that PoC actually tells a very reductive story about computationalism and in this section, before we get to properly analyze the computational aspects of complexity, we will first look at how computationalism evolved in response of the numerous arguments and criticisms addressed to it.

9.3.1 Computationalism and Beyond

From what we said earlier about the theoretical background of DH, it is possible to synthetize all the criticisms against CH in the following two criticisms:

CH treats cognitive systems as sequential machines. While cognition works in real time there is no indication that computational models of cognition may be able of doing the same;
CH treats cognitive systems as they were general-purpose artificial calculator. In fact a cognitive system is a very specialized architecture dedicated to a number of very specialized task. Furthermore, there seems to be no strong relation between the parallel and net-like structure of the brain and the strictly hierarchical architecture of computational devices.

Where both these arguments are really pointing is, in fact, the same direction. What DH accuses computationalism of is to treat cognition as a much simpler phenomenon that it really is. CH is then considered insufficient for effectively explaining the complexity of genuine cognitive phenomena (Piccinini 2010a). However, since the definition of the actual notion of computation has deep consequences on the characteristics of the theoretical framework, some of attention should be dedicated at analyzing this notion and its evolution.

The first accounts of computationalism (Putnam 1967) actually traced a strong analogy between the state of a cognitive system and the state of a turing machine that also survived in computational representational theory of cognition (Fodor 1975). These theories surely suffered from claiming a bit too much about the nature of cognition. While turing machine is currently the best model of computation that we can account for on the general level, it may not be the best model for describing the kind of computations that can be found in the brain. Connectionist models (McLaughlin 2003; Rumelhart and McClelland 1986; McCulloch and Pitts 1943) have been proposed in order to address this "distance" between the kind of digital computer that we use everyday and the kind of distributed machinery that is actually found in real cognitive systems. Through connectionism, a weaker notion of computation is surely achieved. While computations of early models heavily relied on symbolic-representational information processing, in artificial neural networks simply there is no use for the notion of symbol. Further on, ways has been found so that computation may survive the departure from the also influential notion of representation (Piccinini 2006) without having to fall into a contradiction. Even arguments advocating for an analog nature of neural information processing did not rule out the hypothesis that neural computation may well be a case of analog computation (Trautteur 1999), or even a specific case of neural computation (Piccinini and Bahar 2013). Accepting the specificity of the neural substrate of cognition lead to accept that a computational system does not have to be computational at all level (Piccinini 2009). It can well be the case that computational processes may actually emerge from architecture that are only mechanical at the implementation level (Piccinini 2010b) and that, therefore, may even qualify ad dynamical.

From what we have seen, CH does not anymore take specific stance on what may qualify as computational. By doing so CH actually eludes a big deal of the original counter-arguments coming from dynamicist theories. Contemporary computationalism does not need anymore to think at cognitive system as digital computers and CH has become in response nothing more than "the view that intelligent behavior is causally explained by computations performed by the agent's cognitive system (or brain)" (Piccinini 2009, p. 1). Such a broad definition actually points out a very important fact: computation is indeed a flexible and therefore very resilient notion. Much of the explanatory constraints that DH considers unacceptable are derived from the strength of the definition of computation that is taken as fundamental. If in a cognitive system we can find processes that can be modeled in the form of an "effective procedure," then that is now enough to consider that cognitive system as computational. What really is important is not the general definition of computation, but the fact that the cognitively relevant computation is actually found in the brain, which is cognitive per se. It is through this weakening of the definition of computation that complexity makes its way into CH. A more general notion of computation surely is able to accept a wider range of possible implementation while at the same time satisfying the need for a more faithful interpretation of cognitive and neurocognitive facts. However there are also drawbacks. By accepting a notion of computation that kept on becoming weaker and weaker, the actual constraints coming from the implementation level gradually start to become more and more important. In order to consider computational cognitive system in complex way an account should be found for those constraints that come from the neural substrate inside the very notion of neural computation. At the same time also environmental constraints have to be considered in order to reach a complete representation of what it means to be real computational cognitive systems. Complex cognitive systems, even computational one, need to be considered then as intrinsically constrained by ecological and material boundaries. The complexity of the cognitive system may be a useful element of a good computational theory of the mind, but that does not mean that we must consider the fact that even if a computational system is complex enough to implement cognition that does not mean that cognition should not be simple enough to be executed by a computational system.

In light of these facts, in the next paragraph we will consider the actual neuroscientific elements that suggest that there may be limitations at what a particular cognitive systems may be able to do and that there may be a justification for an account of these boundaries in a complex computational theory of cognition.

9.3.2 *Boundaries in the Brain*

If we take now a look at the realization of cognitive systems and, in particular, at the neurophysiological feature of the human brain, we can see that there reason to believe that it does not really show any evidences of being a special piece of machinery. It does not manifests any unique anatomical features in respect of our mammals closest mammal cousins nervous systems (Roth 2012). Furthermore, the human brain seems to be neither the biggest in absolute terms nor in relation to body mass (Jerison 1973, 1991). What, however, surely has is "a relatively thick cortex and a medium neuronal packing density found in hominids, humans have the highest number of cortical neurons found in mammals (and animals), which, however, is only slightly above the numbers found in cetaceans and elephants" (Roth 2012, p. 180). So the specific feature that seems to emerge is the much higher cortical information processing capabilities that comes from high axonal conductional velocity[8] and a short inter-neuronal distance [because of the human particularly high neuronal packing density (Ibidem)]. A feature such as this seems however to indicate at best a performance advantage for the human cognitive system, not supporting the hypothesis humans and their brains have any kind of intrinsically special feature.

So, if the human brain is a biological machine that is characterized by certain performance characteristics, it seems reasonable to ask ourselves if these performances are indeed somehow limited, or if, instead we should continue to the brain as some kind of "impossibility engine" (Cherniak 1990). The first of these options is actually called the "bounded brain hypothesis" (Cherniak 1990; Marois and Ivanoff 2005) and it supports the idea that the human brain only has a limited reservoir of processing resources at its disposal. For these reasons the human brain seems to be constrained by the very characteristics of the its material implementation[9] (Simon 1990; Cherniak 1990), but how can we translate such boundaries in a the computational framework?

9.3.3 Understanding Complexity the Computational Way

At the start of the present section we have seen that contemporary computationalism claims that cognitive processes are computations (Cordeschi and Frixione 2007; Piccinini 2009), hence that a cognitive process always expresses one or more computable function (Massaro and Cowan 1993; Anderson 1990; Cummins 2000). We have also considered how applying complexity to CH leads, similarly to what happened for DH, to considering the cognitive system as a complex, more realistic, and architecturally constrained system. In the previous paragraph we went in search for those constraints and we considered the option that cognitive systems should be

[8]That, on the other hand, comes from the thick myelin sheath that distinguishes the human nervous system (Changizi 2001, 2007).

[9]In particular the relation between the total cortical sheet area and the mean cortical synapse density shows that neurons have at they disposal only a limited space of gray area eligible for connectivity.

considered as bounded. However, when talking in computational terms what we are looking for is a shift from plain computability to a version of it that admits the presence of certain performance boundaries.

Starting from computability it is well known that not every function is Turing-computable,[10] but surely cannot pose too much of an obstacle. Computability checks are, for definition, inherently unbounded in a number of ways, and that is because they rely on an idealized computational model (the Turing Machine) that relies in infinite time and space resources (an infinite tape and infinite computational step). A function is computable in the classical sense, if exists a finite effective procedure that halts after a finite number of steps uses a finite number amount of storage and works for arbitrarily large set of inputs (Enderton 1977). This kind of solution has to be somewhat finite, but there is no a priori indication about how much memory or time does it needs in order to return an output.

As we have seen cognitive agents must be considered as complex cognitive systems and then we have to account for a number of ecological and material boundaries. We have already pointed out how in the real-world decisions have to be taken in split seconds and actions are needed to start and stop at just the right moment. So cognitive processes are bounded at one end by the task that is requested, but they are also limited at the other end by the characteristics of the computational machine (the brain) that executes the computations needed in order to perform that precise task. However, there are no precise estimates of the real computational power of the human brain, but as we have seen only a number of good reasons to believe in the finiteness of our brain capabilities. So it seems reasonable to focus on studying the computational resources requested by those problems that (we believe) our computational brain has to solve in order to express the large array of faculties and phenomena that we usually call mind.

If a cognitive system is computational and it is not trivially considered, then it has to perform not only the right effective procedure given the problem at hand, but it also needs to execute it in the right amount of time and using the right amount of memory in order for it to be cognitively feasible. This is sense in which cognitive computations do not have only to be effective (computable) but also efficient. The efficiency of computation is given by considering its tractability. For this reason, in order to reconsider how cognitive abilities have dealt with limited cognitive resources a "tractable cognition thesis" (van Rooij 2008; Frixione 2001; van Rooij and Wareham 2007) has been proposed. This thesis claims that the mathematical theory of NP-completeness (Garey and Johnson 1979), and computational complexity as a whole, may provide the theoretical constraints needed to account for the boundaries recognized by the bounded brain hypothesis and the need for computationalism to account for complexity. Computational complexity is a theory about the hardness of problems, a theory that studies then the intrinsic complexity of

[10]See, for example, the halting problem (Davis 2004; Turing 1936).

computational task. It analyses how the computational resources (time and memory) required for solving a certain computational problems rise in function of a certain input size[11] (Cook 1971; Goldreich 2008; Arora and Barak 2009; Karp 1972; Steiglitz and Papadimitriou 1982) and establish if certain cognitive problem should be considered as computationally tractable and, then, cognitively feasible. Considering sound this relation between computational tractability and cognitive plausibility it is not a trivial, because while it is true that cognitive systems seems to be somehow bounded at the neural level, there are also cases in which the human cognitive systems actually seem to solve computational non-tractable problems in the blink of an eye (van Rooij 2008, p. 954). So, in order to save the combination of the bounded brain hypothesis and tractable cognition there are two way of proceeding:

We can presume that the information utilized by the cognitive systems is way less that presumed and by doing so reduce our input size, but that may come at the expenses of psychological plausibility (less behavioral cues may imply, in theory, less propositional attitudes);
Or else we can fiddle with the tractability threshold, usually fixed at the polyomial level, in order to admit complex, but also tractable, cognition in conjunction with psychological plausibility.

The second alternative seems more promising and applies a particular account on computational complexity, called parameterized complexity (Downey et al. 1999), in order to devise a refinement of the tractable cognition thesis, the FPT-Cognition thesis[12] (van Rooij 2008). According to this view, the observation that some functions are non-tractable (non-polynomial) only in respect of some small aspects of the input (called the input parameter) may suggest that a number of cognitive capacities that are considered to be computationally non-tractable may be otherwise tractable in respect of a portion of the input size. This may also supports the idea that it is not only the quantity of the input that constraints problem tractability, but also the quality of the input instances and, then, the theoretical and empirical criteria that define the available input contained in the information considered by the cognitive system.

Apart from the actual paradigm being used, what is important for us to note is that this notion of complexity poses a strong accent over the necessity of reaching a cognitively plausible explanation of cognition. Cognitive processes do not have to be only accurately represented in a psychologically plausible way, but they also have to pass a tractability check in order to be considered cognitively plausible. This approach tends toward reaching some kind of equilibrium between our naive experiences of cognition ad what we can reasonably think that our cognitive system may be capable of. In this equilibrium, however, it is the cognitive plausibility criterion that surely has the upper hand.

---

[11]The length of the string representing the input.

[12]That stands for Fixed Paramenter Cognition.

9.4 Conclusion

Ultimately, what should be noted is that the notion of complexity interested both in computationalism and dynamicism shares a commune element. In these theories a complex approach always involves appealing to model of cognition and of cognitive system that shows much more adherence to reality. DH realizes so by making a plea for a more adequate model of explanation that actually encompasses a wider range of possible cognitive phenomena. In dynamicism cognitive phenomena are what makes systems cognitive in the first place, so theories about cognition have to adapt to certain feature of the mind instead of constraining it. CH, on the other hand, reaches a complexity by focusing on a weaker definition for computation and computational systems. However, by doing so, computationalism also has to account for the rising importance of the features and characteristics of the actual neuronal implementation and so it has to introduce certain elements of complexity in its framework in order to demarcate what is relevant for cognition and everything else.

Both these theoretical hypothesis then share the starting point, both of them are in search for a more satisfactory model of cognition. However, it is when complexity steps in that DH and CH part ways. In DH complexity is introduced with a clear emphasis toward psychological plausibility. Under this assumption the best model of cognition is the model that consent to better represent the range of cognitive phenomena usually ascribed to cognitive agents. Releasing cognitive science from the constraints of a certain computationalism makes possible to reach an explanation of cognition that is better because is more psychologically plausible. On the other hand CH uses complexity the other way around. Computationalism had the progressive necessity of weakening its core notion of computation, but instead of embracing the primacy of psychological plausibility it has implemented a notion of complexity with the explicit aim of restricting the range of what we can plausibly call cognitive. Linking computational tractability to cognitive plausibility provides a tool for accurately pinpointing a line of demarcation for distinguishing what is relevant for cognition from what, instead, is only a burden.

So these two approaches at complexity are indeed very different, but it should be worth considering that even if the notions utilized by DH and CH are pointing in different directions they are indeed not contradictory in any sense. Psychological plausibility and cognitive plausibility are both good prerogative of good theories in cognitive science. Maximizing these two theoretical properties requires reaching a point of equilibrium that, on the other hand, would maximize the explanatory power of the particular explanation at hand. The contradiction that started the dispute can be considered genuine only if DH is evaluated in relation to its original computational polemical target. However, such polemical target has gone through an evolution of its core notion so deep that it could be reasonable to argue that the eventual differences between a neural computational system and a cognitive dynamical system may be in fact very small. In conclusion when we take into consideration the notion of complexity entailed by the dynamical and computational approaches, what emerges is a concurrent relation between them that make them complementary in the journey toward a satisfactory theory of cognition.

References

Alligood, K. T., Sauer, T. D., & Yorke, J. A. (1997). Chaos: An introduction to dynamical systems, Textbooks in Mathematical Sciences. Berlin: Springer.

Anderson, J. R. (1990). The adaptive character of thought. Psychology Press.

Arora, S., & Barak, B. (2009). *Computational complexity*. Cambridge: Cambridge University Press.

Byrne, D. (1998). Complexity and postmodernism. *Journal of Artificial Societies and Social Simulation,* vol. 2. London and New York: Routledge.

Changizi, M. A. (2001). Principles underlying Mammalian neocortical scaling. *Biological Cybernetics, 84*(3), 207–215.

Changizi, M. A. (2007). Scaling the brain and its connections: Evolution of nervous systems *3*, 167–180.

Cherniak, C. (1990). The bounded brain: Toward quantitative neuroanatomy. *Journal of Cognitive Neuroscience, 2*(1), 58–68.

Clark, A., & Chalmers, D. (1998). The extended mind. *Analysis* 7–19.

Cook, S. A. (1971). The complexity of theorem-proving procedures. In *Proceedings of the Third Annual ACM Symposium on Theory of Computing* (pp. 151–158). ACM.

Cordeschi, R., & Frixione, M. (2007). Computationalism under attack In M. Marraffa, M. De Caro, & F. Ferretti (Eds.), *Cartographies of the Mind* (pp. 37–49).

Cummins, R. (2000). How does it work? 'Versus' what are the laws?: Two conceptions of psychological explanation. *Explanation and Cognition,* 117–144.

Davis, M. (2004). The undecidable: Basic papers on undecidable propositions, unsolvable problems and computable functions. Courier Dover Publications.

Downey, R. G., Fellows, M. R., & Stege, U. (1999). Parameterized complexity: A framework for systematically confronting computational intractability. In *Contemporary Trends in Discrete Mathematics: From DIMACS and DIMATIA to the Future, AMS-DIMACS Proceedings Series* (vol. 49, pp. 49–99).

Enderton, H. B. (1977). Elements of Recursion theory. *Studies in Logic and the Foundations of Mathematics, 90*, 527–566.

Fodor, J. (1975). The language of thought. New York: Thomas Crowell.

Frixione, M. (2001). Tractable competence. *Minds and Machines, 11*(3), 379–397.

Garey, M. R., & Johnson, D. S. (1979), Computers and intractability (vol. 174). New York: Freeman.

Goldreich, O. (Ed.). (2008). Computational complexity. A conceptual perspetive. Cambridge: Cambridge University Press.

Horgan, T., & Tienson, J. (2006). Cognitive systems as dynamical systems. *The Behavioral and Brain Sciences, 17*(3), 199–204.

Jerison, H. J. (1973). *Evolution of the brain and intelligence*. New York: Academic Press.

Jerison, H. J. (1991). Brain size and the evolution of mind. American Museum of Natural History.

Karp, R. M. (1972). Reducibility among combinatorial problems. Springer.

Marois, R., & Ivanoff, J. (2005). Capacity limits of information processing in the brain. *Trends in Cognitive Sciences*, *9*(6)(June), 296–305.

Massaro, D. W., & Cowan, N. (1993). Information processing models: Microscopes of the mind. *Annual Review of Psychology, 44*(1), 383–425.
McCulloch, W. S., & Pitts, W. (1943). A logical calculus of the ideas immanent in nervous activity. *The Bulletin of Mathematical Biophysics, 5*(4), 115–133.
McLaughlin, B. P. (2003). Computationalism, connectionism, and the philosophy of mind. In The Blackwell (Ed.), *Guide to the Philosophy of Computing and Information*. Cambridge: Blackwell Publishing Ltd.
Penrose, R. (1999). Precis of the emperor's new mind: Concerning computers, minds, and the laws of physics. *Behavioral and Brain Sciences, 13*(04), 643–655.
Petitot, J. (1992). *Physique Du Sens*. Paris: Editions Du CNRS.
Petitot, J. (1999). Naturalizing phenomenology: Issues in contemporary phenomenology and cognitive science. Stanford University Press.
Piccinini, G. (2006). Computation without representation. *Philosophical Studies, 137*(2) (September 21), 205–241.
Piccinini, G. (2009). Computationalism in the philosophy of mind. *Philosophy Compass, 4*(3), 515–532.
Piccinini, G. (2010a). The Resilience of Computationalism. *Philosophy of Science, 77*(5), 852–861.
Piccinini, G. (2010b). Computation in physical systems. In E. N. Zalta (Ed.), *The Stanford Encyclopedia of Philosophy*.
Piccinini, G., & Bahar, S. (2013). Neural computation and the computational theory of cognition. *Cognitive Science, 37*(3), 453–488.
Putnam, H. (1960). Mind and machines. *Mind, Language and Reality*, 362–385.
Putnam, H. (1967). The nature of mental states. *Mind, Language and Reality*, 429–440.
Roth, G. (2012). Is the human brain unique?. In *The Theory of Evolution and Its Impact* (pp. 175–187). Springer.
Rumelhart, D. E., & McClelland, J. L. (1986). *Parallel distributed processing: Exploration in the microstructure of cognition*. Cambridge: MIT Press.
Saltzman, E. L. (1995). Dynamics and coordinate systems in skilled sensorimotor activity. *Mind as Motion: Explorations in the Dynamics of Cognition*, 149–173.
Searle, J. R. (1992). The rediscovery of the mind. MIT press.
Searle, J. R. (1980). Minds, brains, and programs. In J. Haugeland (Ed.), *Behavioral and Brain Sciences* (vol. 3, no. 03, p. 417).
Simon, H. A. (1990). Invariants of human behavior. *Annual Review of Psychology, 41*(1), 1–20.
Steiglitz, K., & Papadimitriou, C. H. (1982). *Combinatorial optimization: Algorithms and complexity*. New Jersey: Printice-Hall.
Thelen, E. (1995). Time-scale dynamics and the development of an embodied cognition. *Mind as Motion: Explorations in the Dynamics of Cognition*, 69–100.
Thom, R. (1988). Esquisse D'une Sémiophysique. Physique Aristotéliciene et Théorie Des Catastrophes. Interéditions, Paris.
Townsend, J. T., & Busemeyer, J. (1995). Dynamic representation of decision-making. *Mind as Motion: Explorations in the Dynamics of Cognition*, 101–120.
Trautteur, G. (1999). Analog computation and the continuum-discrete conundrum. In R. Lupacchini, & E. G. Tamburrini (Eds.), *Grounding Effective Processes in Empirical Laws*. Reflections on the Notion of Algorithm, Bulzoni, Roma.
Turing, A. M. (1936). On computable numbers, with an application to the entscheidungs problem. *Proceedings of the London Mathematical Society, 42*(2), 230–265.
van Gelder, T. (1995). What might cognition be if not computation? *The Journal of Phylosophy, 92*(7), 345–381.
van Gelder, T. (1998). The dynamical hypothesis in cognitive science. *The Behavioral and Brain Sciences, 21*(5), 615–628. (discussion 629–665).

van Gelder, T., & Port, R. (1998). Mind as motion. Exploration in the dynamics of cognition. Cambridge: MIT Press Cambridge.

van Rooij, I. (2008). The tractable cognition thesis. *Cognitive Science: A Multidisciplinary Journal, 32*(6), 939–984.

van Rooij, I., & Wareham, T. (2007). Parameterized complexity in cognitive modeling: Foundations, applications and opportunities. *The Computer Journal*, *51*(3)(November 23), 385–404.

Chapter 10
Language and Brain Complexity

Alessio Plebe and Vivian M. De La Cruz

10.1 Introduction

Humans understand their language thanks to brain processes. While this observation is rather trivial, it is arduous to explain the gap between the bare elements of the brain, like axons, dendrites, ion channels, neurons, and meaning as conveyed by linguistic sentences. It will come as no surprise that this chapter maintains that part of this explanation hinges on complexity. However, unlike several other chapters of this book, our effort is on analyzing complexity in the brain, and in particular, in brain circuits that support language, rather than in language itself, as an abstract entity.

Nevertheless, while steering our attention to neural circuits prior to language, we still share the main inspiration of this book. First of all, our account of "complexity" is essentially mathematical. Even if the use of this term in descriptive and informal senses is common in the biological and neuroscientific communities, in the ranking of animals as "lower" and "higher" according to anatomical features of their nervous system, for example, our mathematical account of complexity is theoretically relevant for the explanatory framework it offers. Here is the second idea that runs across most of this book: the close relationship between complexity and the notion of "emergence."

In its most basic scientific meaning, emergence is the manifestation of a property in a system that lacks counterparts in the properties of the simpler elements it is made of. The relevance of non-obvious properties emerging from local parts of

A. Plebe (✉)
Department of Cognive Science, University of Messina, V. Acicastello 15, 95126 Catania, Italy
e-mail: aplebe@unime.it

V.M. De La Cruz
Department of Educational Sciences, University of Catania, via Teatro Greco 84, 95124 Catania, Italy

F. La Mantia et al. (eds.), *Language in Complexity*,
Lecture Notes in Morphogenesis, DOI 10.1007/978-3-319-29483-4_10

a system is the hallmark of its complexity. Among various behaviors where interacting parts of a system produce emergent phenomena, those collected under the term of "self-organization" have been the most fruitful in providing grounds for the explanation of certain brain processes, such as those supporting linguistic understanding. During self-organization, an initial arrangement of the parts of the system changes in such a way that their interactions become "better" organized, under a certain criteria, and the new "better adapted" organization is observable from outside of the system. Emergence, complexity, and self-organization as well, currently have a wide range of definitions and applications in science, which will briefly be surveyed in Sect. 10.2, before narrowing it down to the notions we will use for these terms in characterizing the brain.

One of the distinctive features of our approach in explaining how, at least in part, the bridge from brain physiology to language relies on self-organization processes is the attempt of adopting a mechanistic view as much as possible. Reconciling mathematical descriptions of phenomena with a mechanistic vision is an approach that has garnered growing consensus inside theoretical neuroscience and philosophy of biology, for its greater explanatory power compared to abstract mathematical formulations. In short, a model can qualify as mechanistic when variables and dependencies among variables in the model correspond to identifiable parts and organizational features of the system. This mapping can be established in a hierarchy of levels, where parts of a system at a higher level have properties that emerge from mechanisms at a lower level. The lowest level we take into consideration has neurons as its simplest parts. Even if the neural cell is certainly a complex system in itself, for the purpose of investigating linguistic processes, we take as our starting point a few basic behaviors of the neuron, such as electrical activity and synaptic plasticity, which will be specified in Sect. 10.2. In the brain, not every phenomenon can be characterized as self-organization, important forms of neural self-organization, however, appear to be essential in scaffolding representations of the external world.

The kind of neural representations we are mostly concerned with is limited to a special part of the brain, the cortex. There are several remarkable features of this structure unique to mammals, that will be briefly reviewed in Sect. 10.3, one of the deepest implications for semantic representations is its ubiquitous organization in two dimensions along its surface. The best solution to representing increasingly complex features of the external world seems to evolve in the cortex through topological structural preservation. The level of *cortical maps* is the next level in our theoretical architecture, for which several mathematical formulations of self-organization are available. We will sketch out in Sect. 10.3.2 the one we adopted in simulating semantic phenomena in the cortex.

Nothing in the human brain, at least according to the current picture provided by neuroscience, and in our account of self-organization driven construction of semantics, can be identified as being unique for language. In the human cortex, even more than in other primates, the majority of neurons are recruited for visual processing, with a relatively small fraction employed for linguistic understanding. Therefore, we lean toward posing a continuity between cortical mechanisms for building structural representations, especially representations of visual objects, and linguistic representations. We will discuss in Sect. 10.4 how mechanistic–mathematical models of cortical maps can simulate the jump from non-linguistic to early linguistic semantic representations of visual objects, and we will discuss how this simulation closely resembles aspects of early language learning in children.

The mapping between sounds produced by caregivers and categories of objects is for infants the first access to language, next comes grasping the idea that utterances can be sequences of different items, and those items combine providing new meaning. One of the most basic sound patterns with a syntactic structure is the adjective–noun sequence, which is known to be one of the first challenges novice language learners face. A mechanistic–mathematical model similar to the previous model discussed, with the addition of a working memory loop, will be presented in Sect. 10.5, and its capacity to simulate the emergence of early syntactic capacities will be considered.

These two models are examples of a research direction labeled as "neurosemantics," the endeavor to explain the development of semantics focusing on the constituent neural processes. In this effort a crucial aspect is the proper understanding of emergent phenomena in the brain, and in the cortex in particular.

10.2 Complexity in the Brain

Complexity of these in last few decades has spread into the heartland of science at large, with a wide range of theoretical frameworks and application domains, for overviews see Scott (2004), Hooker (2010). In the context of brain research, the prevalent use of the notion of complexity has been in grading species according to their repertoire of behaviors and their cognitive abilities (Bullock 2002, 2006; Changizi 2003; Moroz 2009; Shumway 2010). In fact, the whole span of neural machinery, from its first appearance in cnidarians, the early central nervous system of echnoderms, the simple brain in flatworms, to full brains in polychaetes, insects, cephalopods, and vertebrates, rivals anything else in zoology (Roth and Dicke 2013). A tempting straightforward possibility of quantifying brain complexity is in relation to the amount of stored and processed information about the world, and the motor repertoire it is able to perform. As organisms evolve more motor patterns, and require more details on their environment, brains become larger. For Changizi (2003, p. 26) "an entity, or system, is more complex if it can do more kinds of things." However, attempts to relate the number of "kinds of things" an organism can do with macroscopic measures of the brain remain inconclusive. Both the weight and size of the brain in vertebrates scale with body size, in a regular way. The relative brain size as a percentage of body size is also an index with scarce relevance, highest in the mouse and the shrew, and with average values for primates, humans included. Another much-discussed general factor is the encephalization quotient, which indicates the extent to which the brain size of a given species deviates from a sort of standard for the same taxon. This index ranks humans at the top, but is inconsistent with other mammals (Roth and Dicke 2013).

The picture becomes even more complicated when we include invertebrates in the comparison (Chittka and Niven 2009). A honeybee's brain has a volume of about 1 mm^3, yet bees display several cognitive and social abilities previously attributed exclusively to higher vertebrates (Seeley 2003). An alternative view to that of Changizi is held by Bullock (2002, p. 758), which focuses on how the system is made, rather on what it can do. For him complexity "can be defined simply and objectively as the number of distinguishable *kinds of components*." Distinct kinds can be established at different levels, such as varieties of neural cells, or number of distinguishable cortical areas, or also number of different connections to and from source and target brain areas.

Measuring brain complexity by enumerating distinguishable parts at multiple dimensions can probably provide the best guideline when comparing animals of all taxa, but is difficult to assess in practice. When limiting the analysis to mammals, Herculano-Houzel (2009, 2012) found that the number of neurons is a reliable representative of brain complexity. Complexity that is not necessarily a direct correlate of cognitive abilities, for example the elephant brain contains 257 billion neurons, against the 86 billion neurons of the human brain, but as much as 251 billion neurons are found in the cerebellum, and only 5.6 billion in the cortex, against the 16 billion neurons in the human cortex (Herculano-Houzel et al. 2014). Thus, the elephant has a very sophisticated sensorimotor specialization, while humans have more complexity in cognitive behaviors. These different types of evidence supports our choice of taking the neuron as the basic part of the brain system, and of focusing on the cortex as the next level of organization.

10.2.1 Neurons as Computational Devices

Computers, the artifacts managing electricity at a level of sophistication comparable to the brain are made of semiconductors, such as silicon and germanium, not available as organic compounds. Nature's way of dealing with electricity compatible with organic matter is the ion channel. This natural electrical device, whose details have been discovered only recently (Neher and Sakmann 1976), is the precursor of neurons, and appeared 650 million years ago in both ctenophora and early bilateria (Zakon 2012). The first mathematical description of the neural electrotonus in the axon was given by Hodgkin and Huxley (1952), with a system of partial differential equations that reproduce the action potential, the characteristic abrupt electrical impulse, sustained by sodium voltage dependent channels' avalanche effect. Despite its venerable age, the Hodgkin–Huxley model is still the point of reference for the computational simulation of one or just a few neurons, and is embedded in simulators such as GENESIS (Bower and Beeman 1998) and NEURON (Hines and Carnevale 1997). It includes in certain details, the separate effects of the parts (ion channels) involved in action potential phenomena. According to Craver (2007); however, not with a precise enough mapping between variables in the equations and physiological counterparts to be a mechanistic model.

When the focus of modeling is on the interactions between several neurons, not all details of the Hodgkin-Huxley model are necessary, simplified equations have been proposed, such as the integrate-and-fire scheme, where the potential is given by a single linear differential equation with a fixed threshold for the neuron to fire (Gerstner 1999), and the model by Izhikevich (2003).

Independently from the adequacy of description given by the Hodgkin–Huxley model or other similar ones, there are a number of issues that arise when taking neurons as the basic computational devices of the brain: whether they are computational devices at all, and whether they are the most basic ones. The first issue is quite controversial, because it depends on the definition of "computational," which is not so easily defined when applied to objects rather than computers. The controversy is out of the scope of this chapter, but let us at least say that there is a certain convergence on a positive answer, even if the form of computation being performed by neurons might be quite different from that of digital artifacts, and still far from being completely understood (Fresco 2014; Piccinini 2015). The view of the neuron as the basic computational unit of the brain is part of the so-called neural doctrine, proposed by Ramón y Cajal (1899), disputed at large during the past century, but still defended today (Azmitia et al. 2002). A prominent candidate as a more basic unit of computation is the dendrite, which displays certain autonomous computational capabilities (London and Häusser 2005; Sidiropoulou et al. 2006). The main reason for us to keep neurons as the lowest level parts is that neurocomputation built on Cajal's legacy is mature enough to offer tools that are suitable for modeling emergent properties, and allow us to build simulations complex enough to be significant for linguistic semantics. We must keep in mind, however, that the paradigm on which our models are based can change in the future, as the result of better mathematics of the brain, yet to be discovered.

In any case, the mathematical description of the electrical activity in the neuron, or maybe in the dendrite as an autonomous part, is just a small part in the whole story. What is missing is the ability assemblies of neurons have of modifying their structural connectivity based on their own neural activity, which in turn, constructs meaningful computational functions. In humans, most of the organization of the brain at birth is immature, particularly in the cortex, and it is through the continuous interplay between the experience of patterned electrical signals and consequent modifications, that cognitive as well as non cognitive functions mature. Collectively, this process goes under the term of "plasticity," and embraces several mechanisms, such as axon arborization, dendrite rewiring, appearance and disappearance of boutons, and dendritic spines (Møller 2006; Bermúdez-Rattoni 2007). The most relevant mechanism of plasticity involved in daily cognitive business, such as learning a new concept or modifying a previous learned one, is the strengthening or weakening of local synaptic connections between neurons. This is the only form of plasticity in our computational account of semantics from brain complexity. Current mathematical formulations of synaptic plasticity are still influenced by the brilliant intuition of Hebb (1949), formulated long before the possibility of empirically investigating the synapse: it is the coincidence in the timing of the activation of both presynaptic and postsynaptic neurons that produces an increase in the synaptic efficiency. The physiological model of plasticity that most resembles Hebb's law is long-term potentiation (LTP), based on receptors such as NMDA (N-methyl-D-aspartate) (Bliss and Lømo 1973; Bear and Kirkwood 1993). Hebb was right, but he envisioned just one among the many ways of changing synaptic strengths. Long-term depression (LTD) is the converse process to LTP and results in a long lasting decrease in synaptic efficacy (Ito 1989; Crozier et al. 2007). An intriguing and more complex form of plasticity, known as spike-timing-dependent plasticity (STDP) induces synaptic potentiation when action potentials occur in the presynaptic cell a few milliseconds before those in the postsynaptic site, whereas the opposite temporal order results in long-term depression (Levy and Steward 1983; Markram et al. 1997).

10.2.2 How Neural Circuits Come to Represent

Neurons, endowed with the two basic properties of generating action potentials described above, as well as being able to adapt their synaptic connections, assemble brain complexity, and according to Herculano-Houzel, proportionally to their number. While maintaining neurons as parts that build complexity, we move away from the common notion of brain complexity as a measurable ranking of different animals, reviewed at the beginning of this section, and head toward the computational account of complexity, mentioned in the introduction. More precisely, we hold that the brain is complex in that computations across neurons is characterized by emerging properties, chiefly the property of representing objects and events of the world. This claim requires preliminary clarifications on what emerging properties are, and what it is for neurons to represent something.

Historically, emergentism was established in philosophy of science during the late-nineteenth century by a group of British scholars, mostly represented by Broad (1925), as a third alternative in the heated debate between the mechanistic view that living organisms were governed by the same principles of inorganic matter, and the vitalist position. Emergentists confute vitalism, still retaining that the phenomena of life cannot be reduced to the effects of the component substances of the organism, insofar as living matter aggregates at different levels, and each is characterized by certain irreducible properties that "emerge" from lower level properties. Over the years, the impact of emergence declined in philosophy of science, due to the remarkable advances in the understanding of the physical–chemical properties of living matter, but found a renewed vigor these past few decades (Clayton and Davies 2006), in large part thanks to theoretical and mathematical developments. In the 1960s Haken (1978) initiated a research program called "synergetics" to construct an unifying mathematical framework within which quantitative descriptions of self-organizing physical, chemical, and biological systems can be made. The neurocomputational formulations adopted in our models are broadly conceived inside this framework. Other directions of research constituting the current mathematical viewpoint of emergentism include the study of dissipative structures at far from equilibrium conditions, started by Prigogine (1961), or the theory of attractors in chaos theory, first proposed by Ruelle and Takens (1971).

We should add that contemporary emergentism includes positions defending an ontological conception of emergence (Hasker 1999; Silberstein 2006), despite facing known difficulties, such as the downward causation objection (Kim 2006). We are not going to discuss ontological emergentism, since it is transparent to our purposes; the use of a mathematics that is friendly to the epistemological concept of emergence, as a methodological tool in modeling brain complexity.

Neural representation is a debated matter, in part inheriting the range of problems afflicting mental representations in general, such as misrepresentations, how we make mistakes, for example, in categorizing perceptual stimuli (Dretske 1986), or normativity: why we have the clear intuition that in cases of misrepresentation something has gone wrong (for a comprehensive review see Ryder 2009a, b). In addition, neural representations have to fair with their own specific objections. One series of objections comes from defenders of classical computational theory of cognition, such as Ramsey (2007), for whom neurons can only act as "receptors," and as such cannot lead to genuine, "structural," representations. However, as remarked by Sprevak (2011), it is far from and Shagrir (2012) shows how Ramsey's concept of receptors does not meet with the usage of representations in current neural computation (including ours), where structural relations between neural signals and target patterns are common. The idea of neural representations is also challenged from within cognitive neuroscience, by the proponents of the enactive and embodied accounts of cognition (Chemero 2009; Hutto and Myin 2013). One of their flagship arguments is the possibility of dispensing with representations, thanks to dynamic systems theory (Gelder 1995). Here we are not taking up the burden of proving the philosophical consistency of neural representations. We adopt the weak theoretical commitment of Bechtel (2014), who argues that describing neural activity in terms of representation is a useful guiding principle inside mechanistic explanations of brain phenomena.

The general mechanism by which representations emerge in the brain is, in our account, the general ability to detect coincidence in signals, and be affected by them Plebe and De La Cruz (2016, pp. 43–54). The principle of Hebb (1949) provides the lower level, that of two connected neurons, the capacity of capturing the experience of the temporal coincidence of two events, both in the case of external events for sensorial neurons, and of internal signals within the organism. At the highest level, the coincidences between patterns of activation in distant parts of the brain, not just in the same cortical or subcortical area, at a large-scale dimension of signals, correspond to events that are representative at a psychological level. Coincidence detection at brain level shares resemblance with the psychological idea that one fundamental way of acquiring knowledge is by mental association (Ebbinghaus 1885; Thorndike 1892; Shanks 1995; Heyes 2012).

At an intermediate level, coincidence detection involves neurons at such a number that it allows us to compare their activity to an elementary cognitive event, such as a perceptual stimulus. One of the main features of neural organizations at this level is the abundant lateral connectivity, which can always be broken down into single synapses between two neurons, ruled by Hebbian plasticity and the like. However, the signals are now characterized by a high degree of correlation, and the effect of coincidental signals, their self-sustainment, requires an interpretation at a more global level. It can be verified that the ensemble of mechanisms displayed in these assemblies of neurons match with the premises of self-organization, as theorized in synergetics (Haken 1978):

- There if a large number (millions) of elementary elements (neurons) interacting locally;
- Small signal fluctuations might be amplified, this is a direct effect of the non-linear behavior of neurons;
- There is cooperation between fluctuations, in that excitatory lateral connections tend to favor the firing of other connected neurons, and Hebbian law reinforces synapses of neurons that fire frequently in synchrony;
- There is competition as well, in that inhibitory connections can lower the firing rate of groups of cells at the periphery of a dominant active group, and synaptic homeostasis compensates for the gain in contribution from more active cells, by lowering the synaptic efficiency of other afferent cells.

The first attempt to use the mathematical framework of self-organization to describe neural phenomena is due to von der Malsburg (1973), who addressed some of the most important types of organization found in the visual system, essential for building visual representations.

The kind of representation emerging from self-organization of multiple coincidence detection accords well with certain aspects of structural theories of neural representation, such as that proposed by O'Brien and Opie (2004). In their theory, one system structurally "resembles" another when the physical relations among the objects that comprise the first preserve some aspects of the relational organization of the objects that comprise the second. Recently, Nair-Collins (2013) offered a mathematical specification of second-order similarities oriented toward neural representations, as structural "preservation."

In order to characterize neural representations emerging from self-organization, and the kind of "resemblance" or "preservation" they bear with the structures in the external world, we first must focus on the part of the brain that is most relevant for linguistic semantics: the cortex.

10.3 Meaning in the Cortex

In the diversification of the brain throughout evolution, a major turning point occurred about 200 millions years ago, with the formation of a uniform superficial fold, composed of six layers (Striedter 2003) rich of a newly shaped neuron with a pyramidal form, suggestively named "psychic cell" by Ramón y Cajal (1906). It is well agreed upon that the mammalian neocortex is the site of processes enabling higher cognition (Miller et al. 2002; Fuster 2008); however, the reason why the particular way neurons are combined in the cortex makes such a difference with respect to the rest of the brain, remains obscure. There are two striking, and at first sight conflicting, features of the cortex:

- The cortex is remarkably uniform in its anatomical structure and in terms of its neuron-level behavior, with respect to the rest of the brain;
- The cortex has an incredibly vast variety of functions, compared to any other brain structure.

The uniformity of the cortex, with the regular repetition of a six-layered radial profile, has given rise to the search of a unified computational model, able to explain its power and its advantages with respect of other neural circuitry. It is the so-called "canonical microcircuit," proposed in several forms (Shepherd 1988; Douglas et al. 1989; Douglas and Martin 2004). Despite their merits in showing aspects of the cortical structure that may explain its efficiency, such as input amplification by recurrent intracortical connections, any of these circuits could account for a diversity of cortical functions. The answer to how the same circuit could do useful information processing across disparate modalities has to be searched in a different direction: that of plasticity. We reviewed in Sect. 10.2.1 the principles of plasticity at synaptic level, and in Sect. 10.2.2 we sketched how synaptic plasticity offers a key to the general principle of emergent representations through self-organization. Now the same principle is implemented into the kind of organization proper of the cortex.

10.3.1 Topological Representations

If the neuron is the basic element of brain computation, the cortex has its own more powerful computational unit, made up by a collection of cell phenotypes strongly and reciprocally connected inside tiny cylinders called "columns" (Mountcastle 1957; Hubel and Wiesel 1959). Although a precise definition of column, valid all across the cortex, is not straightforward (Horton and Adams 2005) and sometimes confusing (Rakic 2008), it remains a landmark of the cortical organization at its finer degree (Maçarico da Costa and Martin 2010). The next level, relevant for representations, rests on the fundamental concept of "cortical map," as for the concept of column, it has been first introduced by Mountcastle (1957) on the somatosensory cortex, shortly after confirmed by Hubel and Wiesel (1959) in the visual cortex. The two ideas are related: since the vertical dimension is traversed by subcomponents of a unique computational unit, it is along the two-dimensional surface of the cortex that the firing of neurons signals the occurrence of a stimulus on a spot in a sensorial area, and the topological mapping is the first fundamental correlation between sensorial space and cortical space. Cortical maps have been found in nearly all of the sensory and motor areas of the brain (Felleman and van Essen 1991). The most direct kind of topological mapping is of a spatial nature itself, such as the concept of retinotopy, where information represented in the cortical map concerns the topology of the stimulus in the retina. However, as Mountcastle had warned in his early studies, cortical maps should not be interpreted as modified copies of the array of receptors in the periphery. Even if we limit the analysis to a single modality, and the interpretation to the spatial representation of the stimuli, none of the features represented in a cortical map appear to be topographically simple. Maps often contain modular repetitions of small segments of receptor areas, within a global topography (Krubitzer 1995; Vanduffel et al. 2002).

The most intriguing aspect of cortical maps is that ordering in space, within the two dimensions of the cortical surface, can establish a mapping with features of interest in the sensorial stimuli, which are non-spatial in kind. This is the case of the tonotopic organization of the auditory cortex, where the longest spatial dimension of this elongated cortical area is related with the sound frequency, and the shorter dimension with other features of the sound, such as intensity (Verkindt et al. 1995). Also for sensorial stimuli which are spatial in kind, like vision, we found ordering in the two-dimensional space of cortical map which is no mapped to spatial features, like ocular dominance (Wiesel and Hubel 1965) and orientation selectivity (Hubel and Wiesel 1968; Vanduffel et al. 2002). Topological mapping has also been found in the agranular cortex, such as in the posterior parietal cortex, where a mosaic of columns that evoke small, specific hand–forearm movements for reaching and grasping has been observed in monkeys (Kaas et al. 2011). The extent to which the map architecture is ubiquitous as the representation strategy of the cortex, and how map contents should be interpreted, is controversial. Topological order in the homotypical cortex is clearly hard to investigate, to the vast amount of possible dimensions of the relevant stimuli, without any obvious mapping into space dimensions. The suspect is that, given the large amount of direct evidences of topological maps in cortical areas easier to study, higher cortical areas should include many small feature ordered maps too (Kohonen and Hari 2000). Thivierge and Marcus (2007) speculate that topographic maps in the homotypical cortex could be at the basis of abstract reasoning, implementing computations such as universally quantified one-to-one mappings.

Cortical maps, joined with plasticity, complete the picture of the cortex as a representational device, with a kind of organization that may qualify as representation by virtue of its structural similarity property, as in the theoretical proposal of O'Brien and Opie (2004). While they fall short in pointing to brain structures and corresponding resemblance relations, complaining the lack of understanding of the brain, now it is possible to take a step further. Structural resemblance can be established between the topology of the columns in a cortical map, and relations between features of objects in the world, that elicit relevant inputs to that cortical area.

10.3.2 Mathematical Formulations of Self-organization

We already mentioned the work of von der Malsburg (1973) as the pioneering attempt to reproduce neural organization using mathematics within the framework of synergetics. It was based on a system of differential equations, where synaptic efficiency is modified by an amount proportional to the presynaptic and postsynaptic signals, in the case of the coincidence of activity, and periodically all afferent synapsing to the same cortical cell are renormalized, realizing the competition, in that some synapses are increased at the expense of others. Using this mathematical framework, organizations in the visual cortex such as retinotopy, ocular dominance, to orientation sensitivity were reproduced, to some degree, as emergent properties (Willshaw and von der Malsburg 1976; von der Malsburg 1995). Further extensions of his first mathematical approach revealed problematic, the differential equations had no analytic solution, required complex numerical integration and parameters were critical to tune.

A different proposal, offering self-organizing properties at a much cheaper price, in the style of 1980s connectionism, was proposed by Kohonen (1982, 1984, 1995), under the label SOM (*Self-Organizing features Map*). Its mathematical roots are in the vector quantization operation, a method used in signal processing to approximate with a small number of vectors the probability density of a stochastic high dimensional vector (Linde et al. 1980). This algorithm is loosely inspired by the arrangements into two dimensions of the cortical maps, and it was very effective in realizing a relational ordering on input vectors of arbitrary nature (Ritter et al. 1992; Plebe and Anile 2001). The SOM algorithm is a purely phenomenological model, it can successfully reproduce aspects of functional organizations found in neural maps (Yu et al. 2005), but its algorithmic formulation is alien to biological processes.

A formulation that better captures emergent organizations in cortical maps from mechanisms of its parts (neurons) was introduced by Sirosh and Miikkulainen (1997), with the acronym LISSOM (*Laterally Interconnected Synergetically Self-Organizing Map*). It is based on the following mechanisms:

- The intercortical connections of inhibitory and excitatory types, abundant in the cortex (Hunt et al. 2011), that provide feedback and constraints to the overall computation;
- The afferent connections, of thalamic nature, or incoming from lower cortical areas;
- The placement of neurons over two dimensions, as in the cortex;
- The reinforcement of synaptic efficiency by Hebbian learning, the fundamental principle of plasticity;
- The homeostatic compensation of neural excitability, which increasingly appears to be a prominent factor in refining synaptic connectivity (Turrigiano and Nelson 2004).

The LISSOM includes all these features within a static equation, recursive in discrete time steps, thus avoiding the needs of differential equations. It is a simplification with respect to spiking models like GENESIS and NEURON, with the great advantage of allowing the construction of complex hierarchies of cortical maps, inside the recent simulator *Topographica* (Bednar 2009, 2014), built on top of LISSOM.

10.4 Emergence of Visual Object Semantics

The study of the early appearance of language in children, as in the case of names of visual objects, from an emergentist perspective is certainly not new. Recent emergentist approaches to the study of language have grounded their analyses on the frameworks derived from complex adaptive system theories, applying them to language both as an emergent function of the brain and as a CAS in its own right, for a recent collection of views see MacWhinney and O'Grady (2015). These frameworks are: Darwinian evolutionary theory (based on proliferation, competition, and selection), complex systems as structured hierarchically into levels (i.e., properties and structures of the higher levels of complexity are not predictable from those on lower levels), and the theory of timeframes (i.e., processes on different levels are connected to different timescales that come together through competition in the present).

Our account of the emergence of aspects of linguistic meaning is based on the mechanisms of self-organization in the cortex and the topological hierarchical structure it results in a neurocomputational model. As previously mentioned, these mechanisms are characteristic of complex adaptive systems, and demonstrate how from simple local interactions complex structure and processing may emerge. From an early language learning point of view, the models discussed in the sections that follow, indicate how initial categorization abilities, early linguistic knowledge as well as sensitivity to the structure of linguistic input can emerge, thanks to gradual brain maturation and simple mechanisms of associative learning and co-occurance detection, or what we have described in Sect. 10.2.2 as coincidence learning.

10.4.1 Semantics of Visual Objects

Humans are visual creatures. Almost half of our neurons are in fact devoted to the processing of visual signals. The quality of our ophthalmological capabilities is not as good as that of other animals, such as birds, but what we do with what we see, i.e., classify semantically hundreds of object categories based on their visual appearance alone, is pretty interesting. With the models described here, we investigate how the ability to segment and recognize objects on one side, and to segment and recognize words on the other, and the ability to join the two represented entities in linguistic meaning, takes place in the brain at the crossroad between the ventral visual and auditory streams of processing. This is the brain portion simulated in the first model discussed in this section. It is a sort of prelude to a full blown semantics, with a simulation of the full visual pathway that brings light signals into recognition of object categories, together with the auditory pathway, in a simulation of the emergence of the first lexicon, that in infants begins exactly with visual objects. The second model discussed, instead taps into a range of semantic phenomena typically observed in the early stages of language development in children, such as the change in the speed of learning, and the so-called "fast-mapping" phenomenon. Considering how important "seeing" is to our species and the brain power made available to the processing of what we see, it should not be surprising to discover that the first lexicon is highly "object-centric." In fact, the initial vocabulary acquired by very young children is made up to a large extent by nouns referring to visible objects (Bates et al. 1995; Gershkoff-Stowe and Smith 2004). In a cross-linguistic study from 7 different linguistic communities (Argentina, Belgium, France, Israel, Italy, the Republic of Korea, and the United States), Bornstein and RCote (2004) found that children with vocabularies of 51–100 and 101–200 words had more nouns than any other grammatical class. But how does this come to be in a world of competing and often confusing stimuli?

An assorted variety of computational modeling attempts have been made to investigate how the human mind acquires the mapping between words and categories of objects. Some of the very well-known ones belong to the connectionist computational paradigm. For example, Rogers and McClelland (2006) explored the building of simple conceptual systems, using the standard PDP framework (Rumelhart and McClelland 1986). Their model learned by backpropagation, categories such as "flower" and "tree" in correlation with visual features, such as "red" or "branches," together with a fixed set of attributes, such as "can walk," "is living." Despite the higher level of abstraction, and the lack of visual features proper, this model simulated important phenomena such as the emergence of quite general dimensions of similarity without appealing to either physiological or cognitive constraints, but simply as the result of a coherent co-variation of features.

The conceptual system of Rogers and McClelland by itself does not imply a lexical semantics, which instead was the target of the LEX model by Regier (2005). The semantic content in this model, was deprived from any conceptual constituents, most of the focus of the model was on the association with the phonological form of the words. This too was a connectionist model and the level of abstraction was high, with both phonological and semantic features predefined in a conventional way, without any relation to real utterances. A similar approach was pursed by Mayor and Plunkett (2010), whose model we will discuss in the next section since, it explores the same specific aspect of lexical categorization that is the focus of one of our own models.

Other computational models have aimed at understanding visual object recognition, by including in the models a mapping with brain structures (Wallis and Rolls 1997; Deco and Rolls 2004; Rolls and Stringer 2006; Taylor et al. 2005). Not many have extended this mapping beyond the occipital cortex, one example is the model proposed by Kashimori et al. (2007), that includes ITC (Inferior Temporal Cortex) and PFC (PreFrontal Cortex), giving an account of their different roles in categorization. The ITC response is much more influenced by visual features than PFC, even if only by those features important for categorization, and the response in PFC is sustained even after the disappearance of the visual stimulus. Note that the hierarchy of maps of this model reflects the organization of the visual system in monkeys, which differs from that of humans especially in the higher areas. A recent comparison of several models of purely visual categorization is in Khaligh-Razavi and Kriegeskorte (2014). Our model, which we describe below, derives from previous developments on visual object recognition (Plebe and Domenella 2005, 2006, 2007). For what concerns auditory processes, not many neural models have been proposed that simulate them (Näger et al. 2002; Volkmer 2004). This is due to the little that is yet known about the kind of brain computations that lead to word recognition there. The linguistic integration inside our model has been developed in several stages (Plebe 2007b; Plebe et al. 2007, 2010, 2011).

The mechanism of coincidence detection is, as we stated in Sect. 10.2.2, the fundamental principle of cortical representation, and we take it at the core of the emergence of visual objects semantics. Coincidence detection is a multilevel mechanism. At the lower level, it is implemented in local synaptic connections by their sensitivity to the occurrence of simultaneous activation of neighboring units. At an intermediate level, it is responsible for building selectivity in units to recurrent patterns, such as oriented lines in the visual scene, or classes of phonemes. At the highest level, it captures the coincidence of seeing certain objects while hearing the same sound, which then becomes associated with the category of similar objects named by the sound, which is a word.

We will certainly concede that in reality, the formation of object semantics in humans is a complex event, where many different mechanisms converge, for example, Tomasello (1999) underlined the role of pragmatic and social cues in grasping the association between uttered words and objects, something that has been confirmed in several studies (Grassman et al. 2009). A growing amount of new evidence coming from recent developmental studies takes into consideration the child's very own visual perspective, using an embodied approach. These investigations use head cameras mounted on toddlers' heads, and their results indicate that word learning in 18-month olds occurred more efficiently when bottom-up visual information was "clean and uncluttered" (Yu and Smith 2012; Pereira et al. 2014). The optimal viewing conditions were determined by the child's own interaction with objects and with her parents, during the naming event. Our model is not in contrast with these positions, we can assume that social cues drive attentional mechanisms that filter, among all the possible objects presented in a scene, the one focused on by the speaker.

The first model we describe is built upon a number of Topographica maps, simulating several cortical areas, as well as on a few thalamic maps, as shown in Fig. 10.1. The cortical and thalamic areas are the following: lateral geniculated nucleus (LGN), medial geniculated nucleus (MGN), primary visual cortex (V1), secondary visual cortex (V2), ventral occipital (VO), auditory primary cortex (A1), lateral occipital complex (LOC), superior temporal sulcus (STS), prefrontal cortex (PFC).

The structure of the model is based on the accepted perspective that both the visual and the auditory processing pathways in the brain can be divided into two broad main streams, therefore, for the purpose of exploring the emergence of semantics in the brain, it suffices to include only the ventral areas.

Both visual and auditory paths in the model include thalamic nuclei, which perform their own processing of the signals coming from sensory receptors. Given that their maturation is already advanced at the age relevant for the emergence of

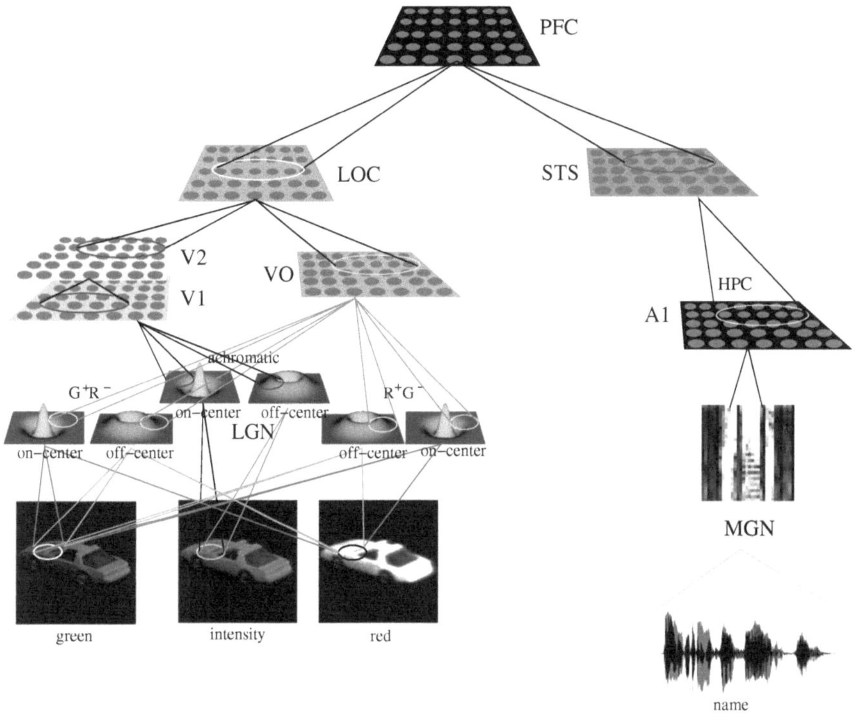

Fig. 10.1 Overall scheme of the object categorization model

semantics, and that their detailed functions are not relevant in the scope of our study, all subcortical processes are hardwired. In the visual path there are separated LGN sheets for the achromatic component, and other sheets for the chromatic components, for the medium and long wavelength. The maps corresponding to early cortical processes are V1, V2, VO, and A1.

The LOC Map, plays a crucial role for object recognition in humans, it is where chromatic and early form processing converge. The highest map in the model, where auditory and visual information meet, is PFC. Certainly the more dramatic simplification of the model is to bind the final semantic processing within a single map. It is well known that the semantic coding of visual objects is spread throughout the brain. However, there is also ample evidence that sustains that the PFC area is deeply engaged in the kind of semantic representation we are interested in investigating (Miller et al. 2002; Freedman et al. 2003; Huey et al. 2006).

10.4.1.1 Experiences and Training

Initially, the model lacks functional processing, with all synaptic connections initialized to small random values. During the experiments the model is exposed to a series of stimuli, which reproduce at a small and essential scale, early human development relevant for its cortical areas. In a first phase only maps V1, VO, and A1 are plastic, and adapt their synaptic connections. The visual stimuli are synthetic random blobs that mimic waves of spontaneous retinal activity that are known to play a fundamental role in the ontogenesis of the visual system (Mastronarde 1983; Katz and Shatz 1996; Thompson 1997).

Those presented to V1 are elongated along random directions, to stimulate orientation selectivity, while blobs to VO are circular, with constant hues, and random size, position, and intensity, in order to induce color constancy. The A1 map is exposed to short trains of waves sweeping linearly around a central frequency, with time duration, central frequencies and sweeping intervals varied randomly.

The second phase involves V2 and STS maps as well. The visual stimuli comprise pairs of elongated blobs, the same previously used for V1, with a coinciding endpoint. These sorts of patterns stimulate the selectivity of units to patterns that are slightly more complex than oriented lines, like corners. The auditory stimuli are synthesized waves of the 7200 most common English words with length in range from 3 to 10 characters, generated using the Festival software (Black and Taylor 1997), tuned at cepstral order 64 and 2.3 s time window.

In a third phase, corresponding to the phase just after eye opening in the newborn, natural images are used. In order to include the identification of an object under different perspectives, the COIL-100 collection was used (Nayar and Murase 1995), where for each of the 100 real childhood related objects, 72 different views are available.

The last phase corresponds to early language learning, and the model is exposed to events in which an object is viewed and a label corresponding to its basic category is heard simultaneously.

The 100 objects viewed by the model were grouped manually into 38 categories. Certain categories, such as "cup" or "medicine" have five exemplars in the object collection, while others, such as "telephone," have only one exemplar. Each category word is converted from text to waves using a male voice, and a female American speaker voice from the Festival software. Both male and female utterances are duplicated at standard and slower speeds.

10.4.1.2 Word Categorization

At the end of the simulated development, several types of topological organization can be found in the maps of the model, which are consistent with the known role of those maps in cortical hierarchy. The V1 map is organized basically with respect to orientation selectivity, in the VO map most units respond to specific hues, regardless of intensity, and the V2 map becomes responsive mainly to angles. Further details on the functions that emerge in the lower areas, not included here, can be found in Plebe and Domenella (2007) concerning V1 and VO, and in Plebe (2007a, 2012) for V2.

The analysis of the higher maps is carried out under the assumption of population coding, the idea that the power of representing information in cortical circuits lies in the combination of the activities of many columnar units. In the current interpretation of population coding, a higher level map may code for a kind of object or fact, and it is the concurrent level of firing of a population of cells in that map that represents a specific instance of the kind. The population of neurons that contribute to the representation of each category in the higher maps is identified by their statistical difference in responding to objects of one category, with respect to objects belonging to all the other categories.

In developing the functions in the STS map, three different sets of stimuli were used: female voices only, male voices only, or using the full set of voices. The population analysis of the final STS map revealed that the accuracy in recognizing names was about 89 % for female or male voices only, and dropped to 66 % in the mixed case, which is still a high value compared with the probability of guessing a name or label by chance, which is of 3 %. The difficuly in the identification of a label when listening to voices of mixed genders is a phenomenon that has been observed in children too. In the final PFC map, the accuracy in recognizing which category an object belongs to is of 88 % in the case of femal voices, of 90 % in the case of male voices, and 70 % in the mixed case. A direct analysis of the final PFC reveals interesting clusterings of information, with linguistic information more clustered on the right side and visual on the left, still with large overlaps. In some cases one modality seems to be prevalent, for example "telephone" is more influenced by the linguistic representation, while for "plug" the visual information is prevailing. It might be due to a richness in phonological form in the first case, while "plug" is phonologically similar to several other names, but singular in shape. In most cases, like in the categories "car" or "fruit" the contribution of the separate linguistic and visual representation is balanced.

10.4.2 Simulations of the Learning of First Words

The model we use to simulate aspects of the emergence of the first lexicon has the components of the previous one, and its overall architecture is similar, but it is applied to a different purpose: the simulation of specific known psycholinguistics phenomena that are typical of lexical acquisition but from a neural perspective. Several particular characteristics have been consistently observed in child development, some refer to the peculiar trend in the speed with which words are learned. The most commonly held view in the literature has referred to this rapid pace in the increase of the child's vocabulary at around 18 months of age, as the "vocabulary spurt" (Lifter and Bloom 1989; Plunkett 1993). However, it has been recast over the years as being more of a gradual linear increase in the child's vocabulary development rather than an "explosion," that can be attributed to a number of factors, such as the child's increasing experience with language as well as to the development of a widening range of cognitive abilities (Elman et al. 1996; Bloom 2000;

Ganger and Brent 2004). A second phenomena is what is known as "fast mapping" (Carey 1978; Dickinson 1984), the ability to grasp aspects of the meaning of a new word on the basis of only a few incidental exposures, gained at around 2 years of age. There is a correlation between the two phenomena, in that fast mapping takes place easier once a critical mass of words have been learned.

Several of the hypotheses proposed to explain these two phenomena, invoke a shift toward mental processes other than those involving development driven by sensorial experience. An idea that spread widely in psychology is rooted in the philosophical notion of "natural kind" (Kripke 1972; Putnam 1975), the very essence of things like animals of chemical substances, that we want our names to refer to, irrespective of their contingent appearance. In psychology this idea has been pushed a step further, in that categories are not just based on a simple evaluation of perceptual similarities, but on a deeper theoretical core, based mainly on causal relationships (Murphy and Medin 1985; Carey and Spelke 1996). There is a large debate on the issue, with many denying the need for an innate core knowledge, for example Tomasello (1999, 2003) defends a perceptual source of linguistic categories, adding the power of social cues, like eye gaze, which help the child in understanding the intended reference of the speaker. A comparison of all the different positions is beyond the current scope, but a deeper discussion is in Plebe et al. (2010), the purpose of the model here presented is to try to assess to what extent sensorial experiences alone can account for the two phenomena at an early stage of development.

The explanation behind our proposed model hinges on the coincidence detection mechanism, mentioned in the previous section. An interesting and similar approach has been recently used by Mayor and Plunkett (2010), with a model based on two SOM maps, one for the visual input and the other for the acoustic input. The units of the two maps are connected, and their efficiencies are updated by increasing those connecting the winning units, as well as the neighboring units on each map. Implicitly, it is an implementation at the top level of the coincidence detection mechanism. The meaning of a word is coded in the cross-connections between the two maps, and is the result of detecting repeated coincidental activation of units responding to object forms and sound forms. The main difference between Mayor and Plunkett's model and ours is in its computational grounding: their model belongs to the connectionist paradigm, and departs from mapping criteria in that the components have no mapping with any particular brain part. These authors argue that this limitation may also be the source of difficulties encountered within their model in learning new words after the visual and auditory maps have stabilized.

The overall scheme of our model is shown in Fig. 10.2, the main difference from the previous model is in the confluence between the visual and auditory streams: not in a LISSOM type map, like those used previously, but a SOM, called ACM (Abstract Categorical Map). Note that in the model of Mayor and Plunkett SOM maps were also used. The main difference is in their placing: in their model the SOM maps act as a substitute to a detailed hierarchy of visual and auditory cortical processes. In our model, the hierarchy is preserved as much as possible, and the SOM is used as an abstract placeholder of the top level mechanism only, which is spread too widely in the brain to be realistically reproduced.

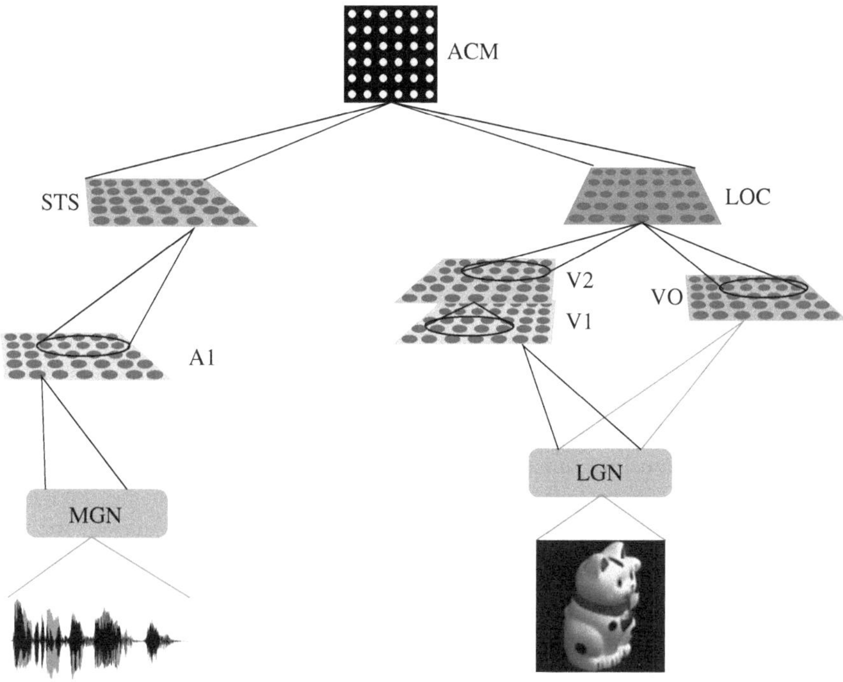

Fig. 10.2 Scheme of the model used in experiments of lexical acquisition

10.4.3 Early Learning Stages

The experiments done with the previous model, involved a single artificial "subject." For the purposes of this model, the experiment is carried out on 500 "subjects," which are clones of the same basic model, but with different experiences. Five stages of lexical learning have been simulated, and in each stage 100 model subjects have been recruited. There is not much difference in the type of experience the models are exposed to compared to the first model we discussed (e.g., elongated blobs input to V1, hue circular blobs for VO, wave trains for A1, pairs of blobs for V2, real images for the visual path, and word waves for the auditory path). The only additions are the neutral combinations of sounds and images, without semantic content, useful to the balancing of the overall number of stimulations. The visual neutral scene is a random image of the Flowers and Landscape McGill collection (http://www.tabby.vision.mcgill.ca/), the auditory neutral pattern is a random musical fragment.

The 500 artificial subjects are individualized from the common model, by generating 500 different subsets from the 100 COIL objects, to which the copies of the model are exposed during the linguistic development phase. The extracted subsets of objects are grouped into five different sizes, corresponding to the five linguistic stages, each with 100 subsets. Although each stage has a fixed number of known objects, because categories of objects in the COIL collection have an uneven number of exemplars, and being that the set of stimuli is selected randomly, the number of known words in a single stage of development varies slightly between individual models. All subsets lack an entire category of objects, as well as a small number of other objects, in order to have samples that are unknown to every artificial subject, to be used as exemplars in the triadic trials. The car category has been chosen because it is composed by a sufficient number of different samples, seven. The use of semantically void samples, aggregating music and landscape images, allows the use of a uniform number of samples for all artificial subjects, avoiding possible artifacts in the results due to the size of the sample set. Since the amodal map in this model is a SOM, it is not possible to apply population coding analysis in interpreting its semantic organization, as done for the PFC map in the previous model. For the SOM a simple labeling operation suffices, in accordance with its winner-take-all principle Kohonen (1995).

Applying the labeling functions, it is now possible to verify the correctness of the identification or the categorization of an object presented to the model, by checking the label of the winner neuron in the ACM map. For example, the performance of the model when presented with an object and its name are evaluated using a labeling equation. If the winner neuron has the same label of the object, the model performed correctly. When presenting an object without naming it, the procedure is the same, but by using another labeling function equation. The overall accuracy is given by the fraction of correct judgments,

10.4.3.1 Fast Mapping Experiments

The experiments simulated the protocol typically used in psycholinguistic experiments for assessing fast mapping. In the training stage an unknown object is presented to the child, labeled with a nonexisting word, the typical utterance is "this is a DAX." In the test stage a small set of objects is presented, and the child is asked to identify the new one: "give me a DAX" (Smith 2001; Regier 2005). The standard used in the experiment is object #23 of the COIL-100 collection, which is a toy red racing car. During the training stage it is shown to the model under 3 different views, very briefly, for 40 training epochs only, as in the fast mapping experiment with children. In the next stage, that of testing, we go about making the model "point" to the chosen object using a mathematical strategy. The ACM map is not evaluated for the winner neuron, as in the ordinary SOM equation. The responses of both new objects at the target unit of the standard object #23, given by another equation, are compared. For most of the objects in COIL-100, taken in turn as the strange object in the tests, all 500 artificial subjects made the correct choice, therefore, the analysis was made by concentrating exclusively on the few objects that confounded the models. The results indicated that all the models showed fast mapping capacities, as well as fast categorization: the models did not have a previous category car, they rapidly acquired it, and the new name behaved as glue for connecting coherent perceptual features of other new objects to this category.

The only four objects that sometimes confounded the model were, at least partially, red. However, we can quickly rule out the hypothesis of color being the dominant feature in establishing a new category, since there are 13 more red objects in the COIL-100 collection that were never chosen as the standard by the model. An important result of the experiments is the progressive improvement in the ability to grasp new categories, at later stages of development. This is far from being a trivial fact for a model, for which "development" just means the dimension of its vocabulary, without any difference in its neural architecture. Further details of these experiments are in Plebe et al. (2010).

From an emergence perspective, what is interesting in the results of the models described in this section is that working within a neurocomputational modeling framework with a minimum of overall structure built in, other than that simplified aspects of what is biologically present in the parts of the brain simulated (maps in visual cortex, simplified auditory cortex, and several thalamic areas) from the local as well as lateral long-range interactions of neuronal activity, not only did an overall organization of the cortical areas emerge, but aspects of linguistic categorization as well. The mechanics of this emergence, hinged on self-organization dynamics, hierarchical processing as well as competition and cooperation between local agents, in this case nodes in lieu of neurons, all representative of the complex adaptive system, that is the brain, and of language as one of its most important emergent functions.

10.5 Emergence of Syntactic Abilities

A long-standing question in language development research concerns the contrast between early word learning and the learning of adjectives. Adjectives, in fact, are almost entirely missing in early productive vocabularies across languages. The ability to acquire adjectives very likely depends on the maturation of brain circuits, especially in the prefrontal cortex. Language development crucially depends, for example, on the maturation of an expanding working memory capacity. The emergence of syntactic processes, such as being sensitive to the order in which words appear, would depend on these enhanced memory circuits found in the temporal-parietal and prefrontal areas, known to develop slowly in ontogeny. This would account for why more complex grammatical forms are acquired later in development: they depend on an expanded memory capacity that is just not available in early infancy. So basically, less memory is necessary for learning nouns initially, but adjective learning is made possible and subsequently easier, only once memory circuits have been enhanced. This section will discuss a neurocomputational model that has in part simulated neural processes behind the learning of adjectives in linguistic development and how an initial sensitivity to word order leading to early syntactic learning or what we refer to as, "embryonic-syntax" might emerge and develop as a result of linguistic exposure and brain maturational processes.

10.5.1 Challenges Involved in the Learning of Adjectives

In contrast to the remarkable rate at which young children learn new nouns, the acquisition of adjectives is sluggish and their use is prone to errors. Even color adjectives, whose meaning seems easily and unambiguously related to perceptual features, seem to be particularly challenging early on.

A number of explanations have been proposed for what seems to be an advantage in learning nouns as opposed to adjectives. Gentner (1978), for example, believes there is a purely logical explanation behind this, for nouns refer to entities, and not relations between entities. For Mintz and Gleitman (2002), the difficulty is in the predicative function of adjectives, in modifying the properties of a concept their learning is necessarily grounded on the acquisition of nouns first. The explanation we propose is instead directly derived from the general mechanism of coincidence detection mentioned earlier in this chapter. An adjective often refers to a single dimension in the space of perceptual features. This is the case, for instance, of adjectives of color, size, and shape. For this reason, the adjective hooks onto a temporal coincidence that is very weak, poorly correlated with a large number of other dimensions in perceptual feature space. It is known that children have difficulty in dealing with more than one perceptual dimension at a time. This kind of difficulty vanishes when learning names of objects, since in this case the sound of the noun of the object category is associated consistently with a rich set of perceptual features. Gradually, with the maturation of the child, mechanisms of selective attention will refine the detection of coincidence between the sound of an adjective, and the single perceptual feature it is regularly correlated with, despite variations in all other dimensions.

There is an important additional aspect of co-occurance or coincidence, and that is that adjectives are rarely heard in isolation. They are most often heard in conjunction with other linguistic items, and therefore represent one of the first experiences with syntax, or word order. Sequences of the type [Adj Noun] are constituents of what could be considered an "embryonic syntax," which departs from the scheme of a holistic relation between a sound sequence and a referenced object in the world. In sum, coincidence detection appears to be an important variable in the process of coming up with a sound trace of the trajectory found in empirical studies on adjective learning (Sandhofer and Smith 2001).

10.5.2 Simulating Working Memory Maturation

From a neurocomputational point of view, the function decoding what we refer to as "embryonic syntax," or the understanding of the sequence (Adj Noun), requires a specific circuital maturation that is simply not available in the early months of life. The prefrontal cortex is involved in this kind of semantic process, and exhibits a trend of development that matches well with the trajectory of adjective learning in children (Aboitiz et al. 2006; Fuster 2001). The prefrontal cortex is considered to be involved in recursive connections with temporo-parietal areas supporting the short-term memory theorized by Baddeley (1992), which would also support phonological decoding. Because the sequence of sounds to be understood, increasingly become longer as the child develops, there is also the increased pressure to temporarily retain them in memory (Vallar and Shallice 2007). The same requirement holds at a higher level, for syntactic units.

The difference in the model we discuss below, with respect to the basic model we used to investigate the emergence of the meaning of words for objects discussed in Sect. 10.4, is in accounting for the maturation of working memory between the prefrontal cortex and the superior temporal sulcus. It is implemented by recursively computing twice the contribution of the auditory pathway in time.

We start off with a basic version of the model (without working memory) such as that seen in Fig. 10.1 in Sect. 10.4. In addition, we will use a "mature" version of this model, where a sort of working memory is simulated. In practice there is a virtual replication of STS into STS', that is a recursive contribution of the same auditory signal delayed in time. The final map PFC will receive contributions from both STS and STS', and therefore can process representations of two words in sequence. The basic form of the model is used as corresponding to early stages in development, at the onset of language learning, around 9–12 months of age, and the complete model at a more mature stage, corresponding to about 14–20 months of age, for additional details on the parameters of the model see Plebe et al. (2013).

10.5.3 Representation of Nouns and Adjectives

The two new models are developed in a manner similar to our previous models, running through prenatal, prelinguistic, and linguistic phases. During the latter, in addition to the naming of categories of objects in coincidence with their view, color adjectives are heard too. From the COIL-100 object collection partitioned into 38 categories, a further partitioning was applied with respect to color, only to those objects sufficiently uniform in hue, and using the 7 basic color categories. In the less mature model, the adjective was heard in isolation, in coincidence with the viewing of one of the objects with the named color. In the complete model, the full sentence with the sequence (Adj Noun) is heard, when seeing an object of category Noun and color Adj.

The analysis of noun and adjective representation is carried out, as with previous versions of the model, using population coding, the statistical assesment of higher maps in establishing the population coding of a category and for evaluating the semantic performance of the model, for further details see Plebe et al. (2013). In testing the immature model (without working memory), the model is presented with either a simultaneous visual appearance of an object and the utterance of its name, or the simultaneous visual appearance of the object and the utterance of its adjective. Tests of the mature model require the simultaneous presentation of a visual object and its noun utterance, followed by the delayed adjective utterance. In the first case the class collects all objects and possible adjectives pertaining to a single noun, while in the second case the class collects all objects and names pertaining to a single adjective. As an alternative, another set of stimuli was used to test ungrammatical sentences, e.g., (Noun Adj)*.

Both models (immature—without working memory, and mature—with working memory) show a good degree of recognition of color adjectives, however, there is a significant improvement when working memory is in place, with an average accuracy of 37 % up to 90 %. The less understood color name in the immature model is "white," with just 18 % accuracy, that increases to 92 % in the mature model. The low value in the immature model is likely due to the vague informative value of "white," compared to other perceptual features of the objects. It is interesting to note that presenting the ungrammatical sentence (Noun Adj)*, the advantage of working memory in the comprehension of adjectives is reduced by half, with an average accuracy of 71 %. Therefore, the model in this version, shows a syntactic selectivity, in that it responds better to sentences where words respect their roles, however, this behavior is not in the form of a norm, in that the violation of the syntax makes the adjective more difficult, but not impossible, to recognize.

We can consider the development of this initial sensitivity to word order as being an emergent quality of the child's increasing linguistic experience in synergy with brain maturation, especially in prefrontal cortical areas, and in particular, in those involved in working memory processes, which are called upon in a variety of higher cognitive tasks, one of which is language.

10.6 Conclusions

Among the plurality of standpoints from which language can be observed and studied, we have taken that of the biological organ that allows humans to understand and use language: the brain. Therefore, we shifted the common thread that crosses this book, complexity, from an intrinsic feature of language as an abstract entity, into a feature of the brain, and in particular a specific feature of the brain in constructing meaning. From brain facts, such as electrical pulses or molecule binding, up to language, the path is very long indeed, often discouraging this perspective of analysis. But it is the same path we travel everyday, without effort, every time we listen, read, chat, or think. It is a path worthy of exploration in order to grasp how human language works, and even if arduous, thanks to the advent of sophisticated neuroscientific research methods and instruments, the possibility of realistically venturing forth, even if with small initial tentative steps, is now a reality.

The research direction that today appears as one of the most promising in explaining cognitive behaviors in terms of brain processes is the mechanistic program. It seems appropriate to tackle brain complexity, by reuniting the reliability of traditional models of reduction, essential for the identification and explanation of brain behavior at the neural cell level, with emergent behaviors at higher levels. This is the approach taken in this chapter to model early language phenomena. The mechanistic approach we adopt has a specific flavor in that it is based on mathematics, formulations of the rules governing interactions between neurons, in a synthesis where the individual components play minor functions, and the overall behavior lies in the way the relations between components self-organize. Even if the models lean upon the abstraction of mathematics, they are still construed as mechanistic, by mapping variables in the equations with physiological counterparts, in particular with the cortical maps engaged in the processes simulated by the models.

Two models have been described: one that simulates the jump from non-linguistic to early linguistic semantic representations of visual objects, in a way that closely resembles early language development in children, and a second that simulates one of the first syntactic forms grasped by children, the sequence of an adjective and a noun. These two models are examples of explorations of the path that takes you from the complex biophysics of the brain up to language, a path today labeled as "neurosemantics," which is still in its infancy, but we deem as one of the most promising for our understanding of human language.

References

Aboitiz, F., Garcia, R. R., Bosman, C., & Brunetti, E. (2006). Cortical memory mechanisms and language origins. *Brain and Language, 98*, 40–56.

Azmitia, E. C., DeFelipe, J., Jones, E. G., Rakic, P., & Ribak, C. E. (Eds.). (2002). Changing views of Cajal's Neuron. Amsterdam: Elsevier.

Baddeley, A. (1992). Working memory. *Science, 255*, 556–559.

Bates, E., Dal, P. S., & Thal, D. (1995). Individual differences and their implications for theories of language development. In P. Fletcher & B. M. Whinney (Eds.), *Handbook of Child Language* (pp. 96–151). Oxford: Basil Blackwell.

Bear, M., & Kirkwood, A. (1993). Neocortical long term potentiation. *Current Opinion in Neurobiology, 3*, 197–202.

Bechtel, W. (2014). Investigating neural representations: The tale of place cells. *Synthese,* 1–35.

Bednar, J. A. (2009). Topographica: Building and analyzing map-level simulations from Python, C/C++, MATLAB, NEST, or NEURON components. *Frontiers in Neuroinformatics, 3*, 8.

Bednar, J. A. (2014). Topographica. In D. Jaeger & R. Jung (Eds.), *Encyclopedia of Computational Neuroscience* (pp. 1–5). Berlin: Springer.

Bermúdez-Rattoni, F. (Ed.). (2007). Neural plasticity and memory: From genes to brain imaging. Boca Raton: CRC Press.

Black, A. W., & Taylor, P. A. (1997). *The festival speech synthesis system: System documentation*, Tech. Rep. HCRC/TR-83. Edinburgh: Human Communciation Research Centre, University of Edinburgh.

Bliss, T., & Lømo, T. (1973). Long-lasting potentiation of synaptic transmission in the dentate area of the anaesthetized rabbit following stimulation of the perforant path. *Journal of Physiology, 232*, 331–356.

Bloom, P. (2000). *How children learn the meanings of words*. Cambridge: MIT Press.

Bornstein, M. H., & RCote, L. (2004). Cross-linguistic analysis of vocabulary in young children: Spanish, dutch, french, hebrew, italian, korean, and american english. *Child Development, 75*, 1115–1139.
Bower, J. M., & Beeman, D. (1998). *The book of GENESIS: Exploring realistic neural models with the general neural simulation system* (2nd ed.). New York: Springer.
Broad, C. D. (1925). The mind and its place. *Nature*. (Kegan Paul, London).
Bullock, T. H. (2002). Grades in neural complexity: How large is the span? *Integrative and Comparative Biology, 42*, 317–329.
Bullock, T. H. (2006). How do brains evolve complexity? An essay. *International Journal of Psychophysiology, 60*, 106–109.
Carey, S. (1978). The child as word learner. In M. Halle, J. Bresnan, & G. Miller (Eds.), *Linguistic theory and psychological reality* (pp. 264–293). Cambridge: MIT Press.
Carey, S., & Spelke, E. (1996). Science and core knowledge. *Journal of Philosophy of Science, 63*, 515–533.
Changizi, M.A. (2003), *The Brain from 25,000 feet—high level explorations of brain complexity, perception, induction and vagueness*. Springer: Berlin.
Chemero, A. (2009). *Radical embodied cognitive science*. Cambridge: MIT Press.
Chittka, L., & Niven, J. (2009). Are bigger brains better? *Current Biology, 19*, 995–1008.
Clayton, P., & Davies, P. (Eds.). (2006). *The re-emergence of emergence: The emergentist hypothesis from science to religion*. Oxford: Oxford University Press.
Craver, C. F. (2007). *Explaining the brain: mechanisms and the mosaic unity of neuroscience*. Oxford: Oxford University Press.
Crozier, R. A., Wang, Y., Liu, C. H., & Bear, M. F. (2007). Deprivation-induced synaptic depression by distinct mechanisms in different layers of mouse visual cortex. In *Proceedings of the Natural Academy of Science USA* (vol. 104, pp. 1383–1388).
Deco, G., & Rolls, E. (2004). A neurodynamical cortical model of visual attention and invariant object recognition. *Vision Research, 44*, 621–642.
Dickinson, D. K. (1984). «First impressions: Children's knowledge of words gained from a single exposure», in. *Applied Psycholinguistics, 5*, 359–373.
Douglas, R. J., & Martin, K. A. (2004). Neuronal circuits of the neocortex. *Annual Review of Neuroscience, 27*, 419–451.
Douglas, R. J., Martin, K. A., & Whitteridge, D. (1989). A canonical microcircuit for neocortex. *Neural Computation, 1*, 480–488.
Dretske, F. I. (1986). Misrepresentation. In R. Bogdan (Ed.), *Belief: Form, Content and Function*. Oxford: Oxford University Press.
Ebbinghaus, H. (1885). *Memory: A contribution to experimental psychology*. New York: Dover.
Elman, J. L., Bates, E., Johnson, M. H., Karmiloff-Smith, A., Parisi, D., & Plunkett, K. (1996). *Rethinking innateness a connectionist perspective on development*. Cambridge: MIT Press.
Felleman, D. J., & van Essen, D. C. (1991). Distributed hierarchical processing in the primate cerebral cortex. *Cerebral Cortex, 1*, 1–47.
Freedman, D. J., Riesenhuber, M., Poggio, T., & Miller, E. K. (2003). Visual categorization and the primate prefrontal cortex. *Neurophysiology and Behavior. Journal of Neurophysiology, 88*, 929–941.
Fresco, N. (2014). *Physical computation and cognitive science*. Berlin: Springer.
Fuster, J. M. (2001). The prefrontal cortex–an update: Time is of the essence. *Neuron, 30*, 319–333.
Fuster, J. M. (2008). *The prefrontal cortex* (4th ed.). New York: Academic Press.
Ganger, J., & Brent, M. R. (2004). Reexamining the vocabulary spurt. *Developmental Psychology, 40*, 621–632.
Gelder, T. V. (1995). What might cognition be, if not computation? *Journal of Phylosophy, 91*, 345–381.
Gentner, D. (1978). On relational meaning: The acquisition of verb meaning. *Cognitive Development, 49*, 988–998.

Gershkoff-Stowe, L., & Smith, L. B. (2004). Shape and the first hundred nouns. *Child Development, 75*, 1098–1114.
Gerstner, W. (1999). Spiking neurons. In W. Maass, & C. M. Bishop (Eds.), *Pulsed Neural Networks*. Cambridge: MIT Press.
Grassman, S., Stracke, M., & Tomasello, M. (2009). Two year olds exclude novel objects as potential referents of novel words based on pragmatics. *Cognition, 112*, 488–493.
Haken, H. (1978). Synergetics—An Introduction, Nonequilibrium Phase Transitions and Self-Organization in Physics, Chemistry and Biology, 2nd edn. Berlin: Springer.
Hasker, W. (1999). *The emergent self*. Ithaca: Cornell University Press.
Hebb, D. O. (1949). *The organization of behavior*. New York: Wiley.
Herculano-Houzel, S. (2009). The human brain in numbers: A linearly scaled-up primate brain. *Frontiers in Human Neuroscience, 3*(Article 31).
Herculano-Houzel, S. (2012). The remarkable, yet not extraordinary, human brain as a scaled-up primate brain and its associated cost. In *Proceedings of the Natural Academy of Science USA* (vol. 109, pp. 10.661–10.668).
Herculano-Houzel, S., de Souza, K. A., Neves, K., Porfirio, J., Messeder, D., & Feijó, L. M., et al. (2014). The elephant brain in numbers. *Frontiers in Nauroanatomy, 8*(Article 46).
Heyes, C. (2012). Simple minds: A qualified defence of associative learning. *Philosophical Transactions of the Royal Society B, 367*, 2695–703.
Hines, M., & Carnevale, N. (1997). The NEURON simulation environment. *Neural Computation, 9*, 1179–1209.
Hodgkin, A. L., & Huxley, A. F. (1952). A quantitative description of ion currents and its applications to conduction and excitation in nerve membranes. *Journal of Physiology, 117*, 500–544.
Hooker, C. (Ed.). (2010). *Philosophy of complex systems*. Handbook of the Philosophy of Science (vol. 10). North Holland, Amsterdam.
Horton, J. C., & Adams, D. L. (2005). The cortical column: A structure without a function. *Philosophical transactions of the Royal Society B, 360*, 837–862.
Hubel, D., & Wiesel, T. (1959). Receptive fields of single neurones in the cat's striate cortex. *Journal of Physiology, 148*, 574–591.
Hubel, D., & Wiesel, T. (1968). Receptive fields and functional architecture of mokey striate cortex. *Journal of Physiology, 195*, 215–243.
Huey, E. D., Krueger, F., & Grafman, J. (2006). Representations in the human prefrontal cortex. *Current Directions in Psychological Science, 15*, 167–171.
Hunt, J. J., Bosking, W. H., & Goodhill, G. J. (2011). Statistical structure of lateral connections in the primary visual cortex. *Neural Systems and Circuits, 1*, 1–12.
Hutto, D. D., & Myin, E. (2013). *Radicalizing enactivism: Basic minds without content*. Cambridge: MIT Press.
Ito, M. (1989). Long-term depression. *Annual Review of Neuroscience, 12*, 85–102.
Izhikevich, E. M. (2003). Simple model of spiking neurons. *IEEE Transactions on Neural Networks, 14*, 1569–1572.
Kaas, J. H., Gharbawie, O. A., & Stepniewska, I. (2011). The organization and evolution of dorsal stream multisensory motor pathways in primates. *Frontiers in Nauroanatomy, 5*, 34.
Kashimori, Y., Ichinose, Y., & Fujita, K. (2007). A functional role of interaction between IT cortex and PF cortex in visual categorization task. *Neurocomputing, 70*, 1813–1818.
Katz, L., & Shatz, C. (1996). Synaptic activity and the construction of cortical circuits. *Science, 274*, 1133–1138.
Khaligh-Razavi, S. M., & Kriegeskorte, N. (2014). Deep supervised, but not unsupervised, models may explain it cortical representation. *PLoS Computational Biology, 10*(e1003), 915.
Kim, J. (2006). Emergence: Core ideas and issues. *Synthese, 151*, 547–559.
Kohonen, T. (1982). Self-organizing formation of topologically correct feature maps. *Biological Cybernetics, 43*, 59–69.
Kohonen, T. (1984). *Self-organization and associative memory*. Berlin: Springer.
Kohonen, T. (1995). *Self-organizing maps*. Berlin: Springer.

Kohonen, T., & Hari, R. (2000). Where the abstract feature maps of the brain might come from. *Trends in Neurosciences, 22*, 135–139.
Kripke, S. A. (1972). Naming and necessity. In D. Davidson & G. H. Harman (Eds.), *Semantics of natural language* (pp. 253–355). Dordrecht: Reidel Publishing Company.
Krubitzer, L. (1995). The organization of neocortex in mammals: are species differences really so different? *Neuroscience, 8*, 408–417.
Levy, W., & Steward, O. (1983). Temporal contiguity requirements for long-term associative potentiation/depression in the hippocampus. *Neuroscience, 8*, 791–797.
Lifter, K., & Bloom, L. (1989). Object knowledge and the emergence of language. *Infant Behavior and Development, 12*, 395–423.
Linde, Y., Buzo, A., & Gray, R. (1980). An algorithm for vector quantizer design. *IEEE Transactions on Communications, 28*, 84–95.
London, M., & Häusser, M. (2005). Dendritic computation. *Annual Review of Neuroscience, 28*, 503–5032.
Maçarico da Costa, N., & Martin, K. A. C. (2010). Whose cortical column would that be? *Frontiers in Nauroanatomy, 4*, 16.
MacWhinney, B., & O'Grady, W. (Eds.). (2015). *The handbook of language emergence*. New York: Wiley.
Markram, H., Lübke, J., Frotscher, M., & Sakmann, B. (1997). Regulation of synaptic efficacy by coincidence of postsyn, aptic APs and EPSPs. *Science, 275*, 213–215.
Mastronarde, D. N. (1983). Correlated firing of retinal ganglion cells: I. Spontaneously active inputs in X- and Y-cells. *Journal of Neuroscience, 14*, 409–441.
Mayor, J., & Plunkett, K. (2010). A neurocomputational account of taxonomic responding and fast mapping in early word learning. *Psychological Review, 117*, 1–31.
Miller, E. K., Freedman, D. J., & Wallis, J. D. (2002). The prefrontal cortex: Categories, concepts and cognition. *Philosophical Transactions: Biological Sciences, 357*, 1123–1136.
Mintz, T. H., & Gleitman, L. R. (2002). Adjectives really do modify nouns: the incremental and restricted nature of early adjective acquisition. *Cognition, 84*, 267–293.
Møller, A. R. (Ed.). (2006). *Neural plasticity and disorders of the nervous system*. Cambridge: Cambridge University Press.
Moroz, L. L. (2009). On the independent origins of complex brains and neurons. *Brain, Behavior and Evolution, 74*, 177–190.
Mountcastle, V. (1957). Modality and topographic properties of single neurons in cats somatic sensory cortex. *Journal of Neurophysiology, 20*, 408–434.
Murphy, G. L., & Medin, D. L. (1985). The role of theories in conceptual coherence. *Psychological Review, 92*, 289–316.
Näger, C., Storck, J., & Deco, G. (2002). Speech recognition with spiking neurons and dynamic synapses: A model motivated by the human auditory pathway. *Neurocomputing, 44–46*, 937–942.
Nair-Collins, M. (2013). Representation in biological systems: Teleofunction, etiology, and structural preservation. In L. Swan (Ed.), *Origins of Mind* (pp. 161–185). New York: Academic Press.
Nayar, S., & Murase, H. (1995). Visual learning and recognition of 3-D object by appearence. *International Journal of Computer Vision, 14*, 5–24.
Neher, E., & Sakmann, B. (1976). Noise analysis of drug induced voltage clamp currents in denervated frog muscle fibers. *Journal of Physiology, 258*, 705–729.
O'Brien, G., & Opie, J. (2004). Notes toward a structuralist theory of mental representation. In H. Clapin, P. Staines, & P. Slezak (Eds.), *Representation in Mind—New Approaches to Mental Representation*. Amsterdam: Elsevier.
Pereira, A. F., Smith, L. B., & Yu, C. (2014). A bottom-up view of toddler word learning. *Psychonomic Bulletin & Review, 2*, 178–185.
Piccinini, G. (2015). *Physical computation: A mechanistic account*. Oxford: Oxford University Press.

Plebe, A. (2007a). A model of angle selectivity development in visual area V2. *Neurocomputing, 70*, 2060–2066.
Plebe, A. (2007b). A neural model of object naming. *Enformatika, 2*, 130–135.
Plebe, A. (2012). A model of the response of visual area V2 to combinations of orientations. *Network: Computation in Neural Systems, 23*, 105–122.
Plebe, A., & Anile, M. (2001). A neural-network-based approach to the double traveling salesman problem. *Neural Computation, 14*(2), 437–471.
Plebe, A., & De La Cruz, V. M. (2016). *Neurosemantics—neural processes and the construction of language meaning*. Berlin: Springer.
Plebe, A., & Domenella, R. G. (2005). The emergence of visual object recognition. In W. Duch, J. Kacprzyk, E. Oja, & S. Zadrony (Eds.), *Artificial Neural Networks—ICANN 2005 15th International Conference, Warsaw* (pp. 507–512). Berlin: Springer.
Plebe, A., & Domenella, R. G. (2006). Early development of visual recognition. *BioSystems, 86*, 63–74.
Plebe, A., & Domenella, R. G. (2007). Object recognition by artificial cortical maps. *Neural Networks, 20*, 763–780.
Plebe, A., De La Cruz, V. M., & Mazzone, M. (2007). Artificial learners of objects and names. In Y. Demiris, B. Scassellati, & D. Mareschal (Eds.), *Proceedings of the 6th International Conference on Development and Learning* (pp. 300–305). IEEE.
Plebe, A., Mazzone, M., & De La Cruz, V. M. (2010). First words learning: A cortical model. *Cognitive Computation, 2*, 217–229.
Plebe, A., Mazzone, M., & De La Cruz, V. M. (2011). A biologically inspired neural model of vision-language integration. *Neural Network World, 21*, 227–249.
Plebe, A., De La Cruz, V. M., & Mazzone, M. (2013). In learning nouns and adjectives remembering matters: A cortical model. In A. Villavincencio, T. Poibeau, A. Korhonen, & A. Alishahi (Eds.), *Cognitive Aspects of Computational Language Acquisition* (pp. 105–129). Berlin: Springer.
Plunkett, K. (1993). Lexical segmentation and vocabulary growth in early language acquisition. *Journal of Child Language, 20*, 43–60.
Prigogine, I. (1961). *Introduction to thermodynamics of irreversible processes*. New York: Interscience.
Putnam, H. (1975). The meaning of "meaning". In *Mind, Language and Reality* (vol. 2). Cambridge: MIT Press.
Rakic, P. (2008). Confusing cortical columns. In *Proceedings of the Natural Academy of Science USA* (vol. 34, pp. 12.099–12.100).
Ramón y Cajal, S. (1899). Textura del sistema nervioso del hombre y de los vertebrados, Vol I, Imprenta y Librería de Nicolás Moya, Madrid, english translation by P. Pasik and T. Pasik, 1997. Berlin: Springer.
Ramón y Cajal, S. (1906). In J. DeFelipe, & E. G. Jones (Eds.), *Cajal on the Cerebral Cortex: an Annotated Translation of the Complete Writings*, Oxford: Oxford University Press, 1988.
Ramsey, W. M. (2007). *Representation reconsidered*. Cambridge: Cambridge University Press.
Regier, T. (2005). The emergence of words: Attentional learning in form and meaning. *Cognitive Science, 29*, 819–865.
Ritter, H., Martinetz, T., & Schulten, K. (1992). *Neural computation and self-organizing maps*. Reading: Addison Wesley.
Rogers, T. T., & McClelland, J. L. (2006). *Semantic cognition—a parallel distributed processing approach*. Cambridge: MIT Press.
Rolls, E. T., & Stringer, S. M. (2006). Invariant visual object recognition: A model, with lighting invariance. *Journal of Physiology Paris, 100*, 43–62.
Roth, G., & Dicke, U. (2013). Evolution of nervous systems and brains. In G. Galizia & P. M. Lledo (Eds.), *Neurosciences—from molecule to behavior* (pp. 19–45). Berlin: Springer.
Ruelle, D., & Takens, F. (1971). On the nature of turbulence. *Communications in Mathematical Physics, 20*, 167–192.

Rumelhart, D. E., & McClelland, J. L. (Eds.). (1986). *Parallel distributed processing: Explorations in the microstructure of cognition*. Cambridge: MIT Press.
Ryder, D. (2009a). Problems of representation I: nature and role. In Symons and Calvo (Eds.), (pp. 233–250).
Ryder, D. (2009b). Problems of representation II: Naturalizing content. In Symons and Calvo (Eds.), (pp. 251–279).
Sandhofer, C. M., & Smith, L. B. (2001). Why children learn color and size words so differently: Evidence from adults' learning of artificial terms. *Journal of Experimental Psychology, 130*, 600–620.
Scott, A. (Ed.). (2004). *Encyclopedia of nonlinear science*. London: Routledge.
Seeley, T. D. (2003). What studies of communication have revealed about the minds of worker honey bees. In T. Kikuchi, N. Azuma, & S. Higashi (Eds.), *Genes, behaviors and evolution of social insects* (pp. 21–33). Sapporo: Hokkaido University Press.
Shagrir, O. (2012). Structural representations and the brain. *British Journal for the Philosophy of Science, 63*, 519–545.
Shanks, D. R. (1995). *The psychology of associative learning*. Cambridge: Cambridge University Press.
Shepherd, G. M. (1988). A basic circuit for cortical organization. In M. S. Gazzaniga (Ed.), *Perspectives on Memory Research* (pp. 93–134). Cambridge: MIT Press.
Shumway, C. A. (2010). The evolution of complex brains and behaviors in african cichlid fishes. *Current Zoology, 56*, 144–156.
Sidiropoulou, K., Pissadaki, E. K., & Poirazi, P. (2006). Inside the brain of a neuron. *EMBO Reports, 7*, 886–892.
Silberstein, M. (2006). In defence of ontological emergence and mental causation. In Clayton and Davies (Eds.), (pp. 203–226).
Sirosh, J., & Miikkulainen, R. (1997). Topographic receptive fields and patterned lateral interaction in a self-organizing model of the primary visual cortex. *Neural Computation, 9*, 577–594.
Smith, L. B. (2001). How domain-general processes may create domain-specific biases. In M. Bowerman, & S. Levinson (Eds.), *Language Acquisition and Conceptual Development*. Cambridge: Cambridge University Press.
Sprevak, M. (2011). William m. ramsey, representation reconsidered. *British Journal for the Philosophy of Science, 62*, 669–675.
Striedter, G. F. (2003). *Principles of brain evolution*. Sunderland: Sinauer Associated.
Symons, J., & Calvo, P. (Eds.). (2009). *The Routledge Companion to Philosophy of Psychology*. London: Routledge.
Taylor, N. R., Hartley, M., & Taylor, J. G. (2005). Coding of objects in low-level visual cortical areas. In W. Duch, J. Kacprzyk, E. Oja, & S. Zadrony (Eds.), *Artificial Neural Networks, ICANN '05, 15th International Conference Proceedings* (pp. 57–63). Berlin: Springer.
Thivierge, J. P., & Marcus, G. F. (2007). The topographic brain: From neural connectivity to cognition. *Trends in Neuroscience, 30*, 251–259.
Thompson, I. (1997). Cortical development: A role for spontaneous activity? *Current Biology, 7*, 324–326.
Thorndike, E. (1892). Animal intelligence: An experimental study of the associative processes. *Animals Psychological Monographs, 2*, 192–205.
Tomasello, M. (1999). *The cultural origins of human cognition*. Cambridge: Harvard University Press.
Tomasello, M. (2003). *Constructing a language: A usage-based theory of language acquisition*. Cambridge: Harvard University Press.
Turrigiano, G. G., & Nelson, S. B. (2004). Homeostatic plasticity in the developing nervous system. *Nature Reviews Neuroscience, 391*, 892–896.
Vallar, G., & Shallice, T. (Eds.). (2007). *Neuropsychological Impairments of Short-Term Memory*. Cambridge: Cambridge University Press.

Vanduffel, W., Tootell, R. B., Schoups, A. A., & Orban, G. A. (2002). The organization of orientation selectivity throughout the macaque visual cortex. *Cerebral Cortex, 12*, 647–662.
Verkindt, C., Bertrand, O., Echallier, F., & Pernier, J. (1995). Tonotopic organization of the human auditory cortex: N100 topography and multiple dipole model analysis. *Electroencephalography and Clinical Neurophisiology, 96*, 143–156.
Volkmer, M. (2004). A pulsed neural network model of spectro-temporal receptive fileds and population coding in auditory cortex. *Neural Computing, 3*, 177–193.
von der Malsburg, C. (1973). Self-organization of orientation sensitive cells in the striate cortex. *Kybernetic, 14*, 85–100.
von der Malsburg, C. (1995). Network self-organization in the ontogenesis of the mammalian visual system. In S. F. Zornetzer, J. Davis, C. Lau, & T. McKenna (Eds.), *An Introduction to Neural and Electronic Networks* (pp. 447–462) (2nd Edn.). New York: Academic Press.
Wallis, G., & Rolls, E. (1997). Invariant face and object recognition in the visual system. *Progress in Neurobiology, 51*, 167–194.
Wiesel, T., & Hubel, D. (1965). Binocular interaction in striate cortex of kittens reared with artificial squint. *Journal of Neurophysiology, 28*, 1041–1059.
Willshaw, D. J., & von der Malsburg, C. (1976). How patterned neural connections can be set up by self-organization. *Proceedings of the Royal Society of London, B194*, 431–445.
Yu, C., & Smith, L. B. (2012). Embodied attention and word learning by toddlers. *Cognition, 125*, 244–262.
Yu, H., Farley, B. J., Jin, D. Z., & Sur, M. (2005). The coordinated mapping of visual space and response features in visual cortex. *Neuron, 47*, 267–280.
Zakon, H. H. (2012). Adaptive evolution of voltage-gated sodium channels: The first 800 million years. In *Proceedings of the Natural Academy of Science USA, 109*, 10.619–10.625.

MIX
Papier aus verantwortungsvollen Quellen
Paper from responsible sources
FSC® C105338

If you have any concerns about our products, you can contact us on
ProductSafety@springernature.com

In case Publisher is established outside the EU, the EU authorized representative is:
Springer Nature Customer Service Center GmbH
Europaplatz 3, 69115 Heidelberg, Germany

Printed by Libri Plureos GmbH
in Hamburg, Germany